Laboratory Investigations on Cold Asphalt Mixes:
A Review

By

OKE, Oluwaseyi Lanre (PhD, Nottingham)
Civil Engineering Department, Faculty of Engineering
Ekiti State University, Ado-Ekiti, Ado-Ekiti, Ekiti State, Nigeria
+2348052682689, oluwaseyi.oke@eksu.edu.ng; seyioke@hotmail.com

1

Table of Contents

List of Figures

List of Tables

1   Overview

Cold mixes with focus on the use of reclaimed asphalt pavements (RAPs) with bitumen emulsions and foamed asphalt are increasingly becoming useful and popular in the construction/rehabilitation of bituminous roads in developed countries of the world. The major benefits of such materials and methods are that they assist in stretching funds available for roads, since they have been acclaimed among others to be economical (old materials are reused), energy and eco-efficient, when used for road construction or upgrading. This book mainly focuses on cold mixes with special interest in RAPs and bitumen emulsion as stabilizing agent.

This book reviews the current laboratory practices usually employed in the design of cold asphalt mixtures. This starts with an explanation on the causes of hardening in bituminous materials and the consequences of such. The review is later extended to laboratory ageing of bituminous materials and cold mixes generally. Researchers have been able to successfully produce RAPs in the laboratory which are typical of those found in developed countries with temperate climates (Bell, 1989; and Scholz, 1995).

2   Ageing in Bitumen and Bituminous Materials Defined

In a review carried out by Airey (2003) on ageing in bitumen and bituminous mixtures, he described the term ageing as hardening which is primarily associated with the loss of volatile components and oxidation of the bitumen during asphalt mixture construction (short-term ageing) and progressive oxidation of the in-place material in the field (long-term ageing). Scholz (1995) in his work, defined bitumen ageing as hardening in bitumen as a result of compositional changes in the bitumen. Srivastava and Rooijen (2000) considered that the phenomenon in which the rheological properties of bitumen change with time (i.e. bitumen becomes harder and more elastic) is referred to as ageing. Notwithstanding the associated causes, ageing in bitumen could be defined simply as hardening due to changes in molecular structure and/or chemistry of bitumen with a consequential effect on its rheology.

3   The Cause, Nature and Process of Ageing in Bitumen and Bituminous Materials

Activities tailored towards understanding ageing in bitumen have a relatively long history. Welborn (1979) recorded that since the 1930s, research has continued to develop an understanding of the factors contributing to short term and long term ageing. Scholz (1995) reported that many researchers have investigated age hardening of bitumens and bituminous mixtures and have provided significant advances toward a better understanding of the mechanisms of age hardening. In an extensive investigation, Traxler (1963) listed 15 factors (see Table 1) that could aid ageing in bitumen and which consequently lead to a change in its chemical, rheological and adhesion characteristics. Mastrofini and Scarsella (1999) on the same issue opined that, colloidal structure and stability of asphaltenes in residues and bitumens are directly connected with ageing. Srivastava and Rooijen (2000) averred that some aggregates act as catalysts for the oxidation reactions, while others have inhibitive effects and that during service life, oxidation depends, apart from climatic conditions and the ageing resistance of the bitumen, mainly on the amount of air voids in the asphalt and the bitumen film thickness.

Lu and Isacsson (2002) opined that bitumen ageing mechanisms in pavements are two in nature with the main being irreversible, and normally characterised by chemical changes of the binder which in turn have an impact on the rheological properties of the bitumen. They identified oxidation, loss of volatile components and exudation as the factors responsible for the mechanism. On the other hand, they stated that the second mechanism is a reversible process which is called physical hardening (steric hardening). Karlsson (2002) from his experience found out that the degree of ageing depends on temperature, air void content of the mixture and chemical composition of the binder. Airey (2003) advanced that molecular structuring over time (steric hardening) and actinic light (primarily ultraviolet radiation, particularly in desert conditions) may also contribute to ageing. Judycki and Jaskula (2003) in line with Airey's findings declared in their work that short term ageing of bitumen occurs during bitumen storage, mix production and laying; while long term ageing occurs during the service life of the asphalt pavement. Mouillet et al (2003) simply stated that ageing in conventional bitumen is a complex process. In a more recent work, Chen et al (2007) corroborated the findings of earlier researchers by reporting that the degree of ageing in bitumen depends on temperature, air void content of the mixture and the chemical composition of the binder, and that during bitumen oxidative ageing, the saturates remain the same while solubilising aromatics decrease in quantity, because the aromatics react with oxygen to produce asphaltene, which causes the asphaltene content to increase. Also, Shen et al (2007) in agreement with Scholz's findings, using gel permeation chromatography (GPC) showed that compositional changes of the blends of aged binders are reflected well by the chromatographic profiles measured by the GPC. Meanwhile, Carswell et al (2008) in a general review commented that durability of asphalt pavements is a major issue and that it is a fairly complex problem because it involves a number of parameters, with binder ageing and moisture damage considered as prominent.

Table 1: Factors which Affect Chemical Rheological and Adhesion Characteristics of Bitumen (Traxler, 1963)

Factors	Influenced by					Occurs	
	Time	Heat	Oxygen	Sun-light	Beta & Gamma Rays	At Surface	In Mass
Oxidation (in dark)	√	√	√			√	√
Photo oxidation (direct light)	√	√	√	√		√	
Volatilisation	√	√				√	√
Photo oxidation (reflected light)	√	√	√	√		√	
Photo chemical (direct light)	√	√		√		√	
Photo chemical (reflected light)	√	√		√		√	
Polymerisation	√	√		√		√	√
Development of internal structure (Ageing, Thixotropy)	√					√	√
Exudation of Oil (Syneresis)	√	√				√	
Changes by nuclear energy	√	√			√	√	√
Action by water	√	√	√	√		√	
Absorption by solid	√	√				√	√
Absorption of components at solid surface	√	√				√	
Chemical reactions or catalytic effects at interface	√	√				√	√
Microbiological deterioration	√	√	√			√	√

From the forgoing, it could be conveniently summarised that ageing in bitumen and bituminous mixtures is a complex phenomenon that ultimately manifests as hardening of the binder. Airey (2003) and Bell (1989) from their extensive literature reviews, both agreed that the majority of researchers considering ageing of bitumen and bituminous mixtures have come to the conclusion that, (a) the loss of oily components by volatility or absorption, (b) changes in composition by reaction with atmospheric oxygen and (c) molecular structuring that produces thixotropic effects are the major factors responsible for ageing. Thus, most researchers have limited their investigations to these three factors and indeed recent research works have been towing this line as will be discovered in the subsequent sections of this book.

4   Consequences of Ageing in Bitumen and Bituminous Mixtures

Short term ageing of bituminous mixtures i.e. during mixing, transportation, laying and compaction, is most desired by pavement engineers for the beneficial role of facilitating the necessary gain in stiffness required for pavements to support traffic after construction. However, when the ageing becomes excessive, it could lead to a failure of the pavement functionally and structurally. Scholz (1995) reported that excessive age hardening can result in a brittle bitumen- one with significantly reduced flow capabilities- which contributes to various forms of cracking in the bituminous mixture. In an earlier work, Kim and Burati (1993) discovered that ageing of bitumen either in the field or in the laboratory results in drastic changes in molecular size that leads to changes in the consistency of the bitumen. Airey (1997) from his findings stated the consequences of bitumen ageing as, increase in complex modulus with corresponding decrease in phase angle (i.e. showing an increase in elastic behaviour of bitumen). In considering pavements in the tropics, Smith and Jones (1998) observed that ageing causes premature top down cracking in pavements in the tropics. In buttressing this fact further, Mastrofini and Scarsella (1999) in their work agreed that bitumen ageing is a very complex process resulting in hardening of bitumens and embrittlement, both in application and in service, which contributes greatly to the deterioration of paving applications.

Meanwhile, Soenen et al (2000) in their studies on polymer modified bitumen confirmed that, due to ageing, the bitumen increases its stiffness and elasticity, and that the interaction between aged polymer and aged bitumen is different from the un-aged state. Srivastava and Rooijen (2000) on another note argued that hot temperature induces ageing which results in permanent deformation and surface cracking of asphalt mixtures. Lu and Isacsson (2002) opined that bitumen ageing is one of the principal factors causing the deterioration of asphalt pavements. They reported that ageing influences bitumen chemistry and rheology significantly and that chemical changes such as, formation of carbonyl compounds and sulfoxides, transformation of generic fractions (asphaltenes, resins, saturates and aromatics), and increases in amount of large molecules, molecular weight and polydispersity take place during ageing (see Figure 1 and Table 2). Karlsson (2002) stated that ageing in principle has negligible effect on the rate of diffusion in aged binder since the diffusion only takes place in the maltene phase which is relatively unaffected by ageing. On the other hand, Widyatmoko (2002) reported that the capacity for healing reduces more rapidly after ageing for semi-blown binders. Airey (2003) submitted that ageing of the bituminous binder is manifested as an increase in its stiffness (or viscosity see Figure 2) which consequently results in the stiffening of the mixture. Section 5, Volume 7 of the 1999 version of the UK Design Manual for Roads and Bridges, opined that conventional binders, as thin films on the aggregate particles, age in the presence of air leading to fretting and ravelling (loss of aggregate in the surface), cracking, and finally to failure. It was further stated that the rate of change in the binder depends on the voids in the mixture.

Bearsley et al (2004) in their studies reported that the structure and fluorescence of the asphaltene phase does not appear to alter radically upon oxidative ageing.

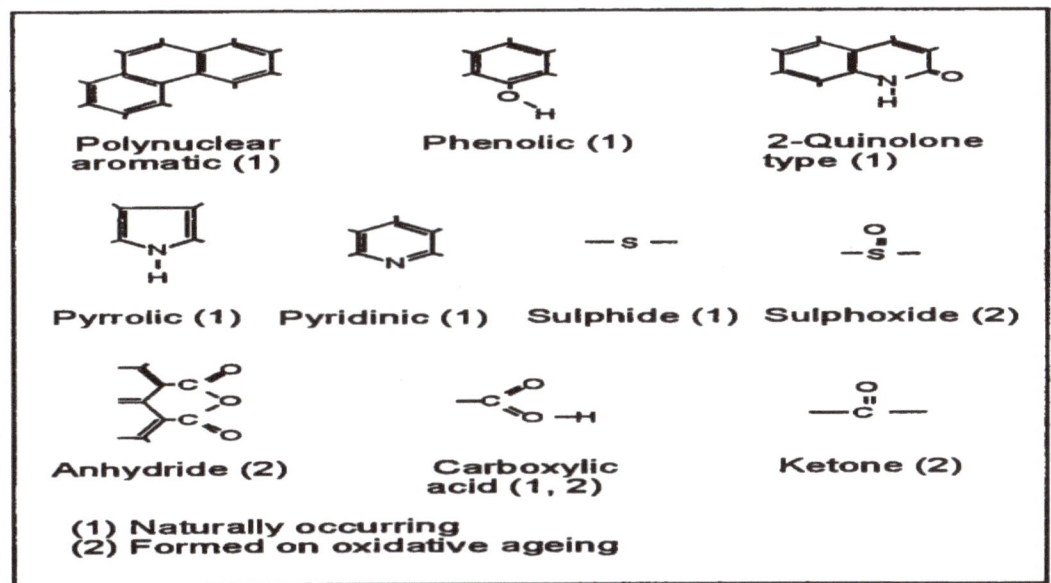

Figure 1: Chemical Functionalities in Bitumen Molecules Normally Present or Formed on Oxidative Ageing (Petersen, 1984)

Table 2: Chemical Functional Groups Formed in Bitumens During Oxidative Ageing (Petersen et al, 1974)

Bitumen	Concentration (moles per litre)				Average Hardening Index[b]
	Ketone	Anhydride	Carboxylic Acid[a]	Sulphoxide	
B-2959	0.50	0.014	0.008	0.30	38.0
B-3036	0.55	0.015	0.005	0.29	27.0
B-3051	0.58	0.020	0.009	0.29	132.0
B-3602	0.77	0.043	0.005	0.18	30.0

Note: Column oxidation, 130ºC, 24hours, 15μm film
[a] Naturally occurring acids have been subtracted from reported value.
[b] Ratio of viscosity after oxidative ageing to viscosity before oxidative ageing.

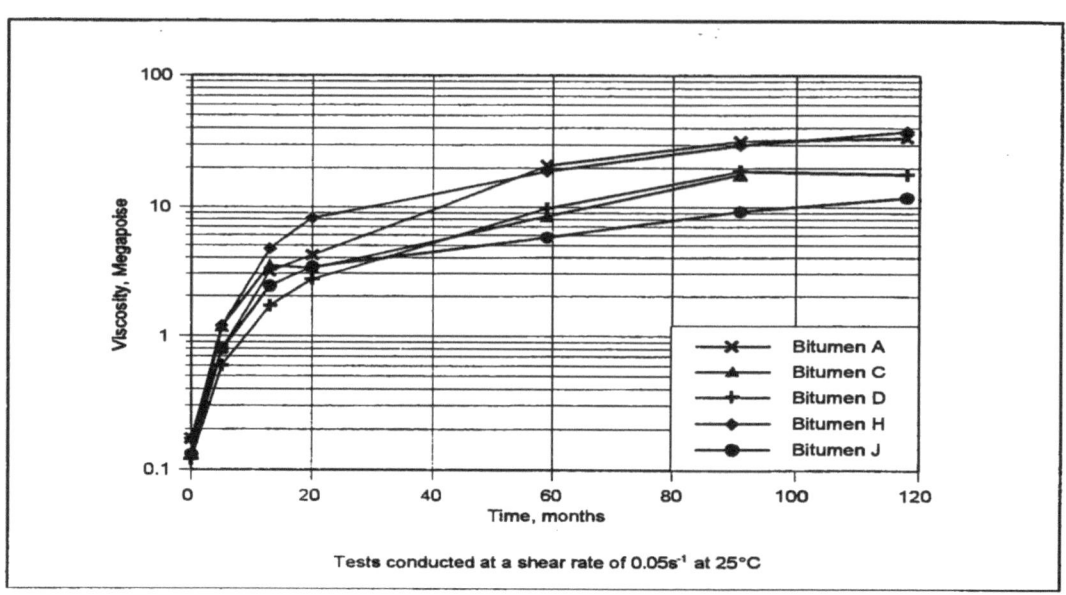

Tests conducted at a shear rate of 0.05s⁻¹ at 25°C

**Figure 2: Viscosity Change of several Bitumens During
Service in Pavements (Zube and Skog, 1969)**

Walubita et al (2005) reported that, ageing reduced HMAC mixture fatigue resistance and its ability to heal and that, ageing plays a significant role in HMAC mixture fatigue performance and subsequently advised that it should be incorporated in fatigue design and analysis. Airey et al (2005) are of the opinion that, age hardening can have two effects, either increasing the load bearing capacity and permanent deformation resistance of the pavement by producing a stiffer material or reducing pavement flexibility resulting in the formation of cracks with the possibility of total failure. Lately, Romera et al (2006) argued that asphaltene contents increase with ageing and that aged bitumen needs a high mixing temperature (>200°C) to behave like a fluid material able to wet, adhere and envelop aggregate.

It is thus obvious from the account of previous researchers that, all things being equal, ageing in bituminous mixtures could play a beneficial role most especially when it is short term in nature as it assists the asphalt layer in a flexible pavement to support the designed traffic loads, since it moderately stiffens the asphalt layer. On the contrary, when the asphalt layer is aged severely or excessively, it consequently leads to significant hardening, which imposes some undesirable attributes on the asphalt layer in which its ability to flex gets reduced, and thus the pavement assumes a position in which it is at the risk of imminent failure by cracking since the fatigue life has reduced drastically. Ageing too in this regard, could cause permanent deformation, ravelling and fretting among others. All these

10

eventually result in surface defects which ultimately lead to the total failure of the road if un-attended to.

5 Laboratory Protocols that Simulate Field Ageing in Bituminous Road Pavements

The need for laboratory simulations generally to mimic the behaviour of engineering structures while in service cannot be overemphasized. Laboratory simulations serve as avenues through which the performances of such structures under a near to service/field condition are evaluated before being finally put in place so that such structures would eventually serve their intended design functions. Moreover, a pavement which is the structure being considered here, involves a huge investment, thus every effort at the disposal of road builders must be harnessed to ensure that such roads perform their roles efficiently and effectively and also meet the design life at moderate costs. To this end, lots of simulations have been carried out and are still ongoing to ascertain the behaviour/performance of asphalt pavements while in service. One such is the effect of ageing on the performances of bitumen and bituminous mixtures. Mastrofini and Scarsella (1999) observed that quite a number of researchers have tried to simulate in the laboratory the ageing that takes place in service on asphalt pavements in order to foresee bitumen behaviour during application and service life. Notable among these are the works carried out by Bell (1989) which was sponsored under the Strategic Highway Research Program (SHRP) of the National Research Council in the USA, and Scholz (1995) which was carried out under the BITUTEST programme in the UK. In his work, Bell (1989) summarised the various methods that have been used to simulate field ageing in binders and bituminous mixtures. Similarly, Airey (2003) did an appraisal on the different laboratory methods that have been advanced by researchers for simulating ageing that occurs in the field in asphaltic pavements both in the short and long term. Indeed, these works showed that there are a lot of methods being used. Some of these methods are summarised in Tables 3.3 and 3.4.

Ever since the works of Bell and Airey, a number of simulations have been conducted by researchers in this regard. For example, Mouillet et al (2003) developed a new ageing protocol in which ageing of sample was done by the simulated oxidation of Polymer Modified Bitumens (PmBs) in the ageing cell, by heating the sample at 130°C for 2 hours under synthetic air (80% N_2, 20% O_2). Before this, the cell was heated under neutral gas from 25°C to 130°C at 111°C/min heating rate. They eventually concluded that the method used for simulating ageing in their work would be useful for other kinds of organic materials for which a precise ageing knowledge is essential.

In another elaborate work, Hachiya et al (2003) discovered that asphalt concretes seem to become hard and brittle owing to both heat ageing and exposure to natural conditions. They also confirmed that ageing of asphalt in natural conditions varies with depth below the surface. They noted that asphalt near the surface hardens to a greater extent and that flexural strength changes fairly little with ageing time, strain at failure decreases clearly with ageing time and asphalt concrete mixture increases in stiffness with ageing time. They further remarked that the change in composition of bitumen due to ageing is a decrease in aromatics and an increase in asphaltenes and that however, heat ageing merely causes hardening. They also found out that samples from oxidative ageing of asphalt concrete had flexural strength and stiffness that increased with the duration of accelerated oxygen ageing. These authors opined that as asphalt cannot be aged in this procedure as well as in outdoor exposure, a further modification such as a combination of heat ageing and oxygen should be adopted to simulate actual ageing. The work ended by concluding that ageing rate of asphalt concrete depends upon various factors such as asphalt type, asphalt content, aggregate, and environment. They also remarked that ageing is more pronounced in the top 5mm layer of the pavement than the other parts of the pavement, thus ageing decreases with depth, and that even if heat ageing could be used to evaluate hardening of asphalt concretes in terms of mechanical properties, it cannot be used to evaluate the changes in physical and chemical properties of asphalt.

On the same note, Walubita et al (2005) used three ageing protocols in their study. The exposure conditions lasted for 0, 3, and 6months at 60°C, simulating up to approximately 12 years of Texas field Hot Mix Asphaltic Concrete (HMAC) at the critical pavement service temperature with no explanation on what critical meant. All loose HMAC mixtures were subjected to the standard AASHTO PP24hr short-oven ageing process at 135°C prior to 60°C ageing. After HMAC mixture testing, aged binders were extracted for testing to characterize the binder's chemical and physical properties. They found out that the mixture tensile test conducted at 20°C indicated that as HMAC ages, it becomes more brittle, thus breaking under tensile loading at a lower strain level. Indeed the results of these simulations are instructive. Bell et al (1994) reported that two days of long term oven ageing at 85°C is representative of pavement up to 5 years old depending on climate. Four days of oven ageing at 85°C appears to be representative of field ageing of about 15yrs in a Wet-No Freeze zone and about 7yrs in a Dry Freeze zone.

Table 3: Bitumen Ageing Methods (Airey, 2003)

Test method	Temperature (°C)	Duration	Sample size (g)	Film thickness	Extra features
Thin film oven test (TFOT) (Lewis and Welborn, 1940)—ASTM D1754, EN 12607-2	163	5 h	50	3.2 mm	–
Modified thin film oven test (MTFOT) (Edler et al., 1985)	163	24 h	–	100 μm	–
Rolling thin film oven test (RTFOT) (Hveem et al., 1963)—AASHTO T240, ASTM D2872, EN12607-1	163	75 m	35	1.25 mm	Air flow—4000 ml/min
Extended rolling thin film oven test (ERTFOT) (Edler et al., 1985)	163	8 h	35	1.25 mm	Air flow—4000 ml/min
Nitrogen rolling thin film oven test (NRTFOT) (Parmeggiani, 2000)	163	75 m	35	1.25 mm	N_2 flow—4000 ml/min
Rotating Flask Test (RFT)—DIN 52016, EN12607-3	165	150 m	100	–	Flask rotation—20 rpm
Shell microfilm test (Griffin et al., 1955)	107	2 h	–	5 μm	–
Modified Shell microfilm test (Hveem et al., 1963)	99	24 h	–	20 μm	–
Modified Shell microfilm test (Traxler, 1961; Halstead and Zenewitz, 1961)	107	2 h	–	15 μm	–
Rolling microfilm oven test (RMFOT) (Schmidt and Santucci, 1969)	99	24 h	0.5	20 μm	Benzene solvent
Modified RMFOT (Schmidt, 1973)	99	48 h	0.5	20 μm	1.04 mm φ opening
Tilt-oven durability test (TODT) (Kemp and Prodoehl, 1981)	113	168 h	35	1.25 mm	–
Alternative TODT (McHattie, 1983)	115	100 h	35	1.25 mm	–
Thin film accelerated ageing test (TFAAT) (Petersen, 1989)	130 or 113	24 or 72 h	4	160 μm	3 mm φ opening
Modified rolling thin film oven test (RTFOTM) (Bahia et al., 1998)	163	75 m	35	1.25 mm	Steel rods
Iowa durability test (IDT) (Lee, 1973)	65	1000 h	TFOT residue—50	3.2 mm	2.07 MPa—pure oxygen
Pressure oxidation bomb (POB) (Edler et al., 1985)	65	96 h	ERTFOT residue	30 μm	2.07 MPa—pure oxygen
Accelerated ageing test device/Rotating cylinder ageing test (RCAT) (Verhasselt and Choquet, 1991)	70–110	144 h	500	2 mm	4–5 l/h—pure oxygen
Pressure ageing vessel (PAV) (Christensen and Anderson, 1992)	90–110	20 h	RTFOT or TFOT residue—50	3.2 mm	2.07 MPa—air
High pressure ageing test (HiPAT) (Hayton et al., 1999)	85	65 h	RTFOT residue—50	3.2 mm	2.07 MPa—air

Table 4: Bituminous Mixture Ageing Methods (Airey, 2003)

Test method	Temp (^{0}C)	Duration	Sample Size/Condition	Extra Features
Production Ageing (Von Quintas, 1988)	135	8,16,24,36h	Loose material	-
SHRP short-term oven ageing (STOA)	135	4h	Loose material	-
Bitutest protocol (Scholz, 1995)	135	2h	Loose material	-
Ottawa sand mixtures (Paul and Welborn, 1952)	163	Various periods	50 x 50mm^2 cylinders	-
Plancher et al (1976)	150	5h	250 x 40mm$^2\Phi$	-
Ottawa sand mixtures (Kemp and Prodoehl, 1981)	60	120h	-	-
Hugo and Kennedy (1985)	100	4 or 7 days	-	80% RH
Long-term ageing (Von Quintas, 1988)	60	2 days	Compacted Specimens	-
	107	3 days		
SHRP long –term oven ageing (LTOA)	85	5 days	Compacted Specimens	-
Bitutest protocol (Scholz, 1995)	85	5 days	Compacted Specimens	-
Kumar and Goetz (1977)	60	1, 2, 4, 6, 10 days	Compacted Specimens	Ait at 0.5mm of water
Long-term ageing (Von Quintas, 1988)	60	5 to 10 days	Compacted Specimens	0.7MPa -air
Oregon mixtures (Kim et al, 1986)	60	0,1 ,2, 3, 5 days	Compacted Specimens	0.7MPa -air
SHRP low pressure oxidation (LPO)	60 or 85	5 days	Compacted Specimens	Oxygen-1.9l/min
Khalid and Walsh (2000)	60	Up to 25 days	Compacted Specimens	Air-3l/min
PAV mixtures (Korsgaard, 1996)	100	72h	Compacted Specimens	20.7MPa -air

Oven ageing at 100°C for one, two and four days causes similar changes in modulus to 85°C ageing for two, four and eight days, but damages the samples in the process. Oven ageing at 85°C is thus considered to be more reliable though nothing was mentioned about the nature of the damage caused by oven ageing at 100°C. In line with these assertions, Hinds (2007) in his analysis, stated that, using the SHRP long term oven ageing technique reflects an effect equivalent to 15years in a wet no freeze climate. Kulash (1994) had earlier reported that, in the Pressure Ageing Vessel, bitumen aged at 2.07MPa air for 20 hours at 100°C simulates a 5year aged level in a pavement in the temperate region, and that the forced draft oven is able to simulate field ageing (short and long term). Airey et al (2005) observed that the Saturation Ageing Tensile Stiffness Test (SATS) was able to reproduce the 60% reduction in stiffness modulus found for the acidic aggregate HMB mixture in the field.

It could be summarised from the foregoing that, ageing tests are basically two i.e. those carried out on binders and those carried out on bituminous mixtures. Airey (2003) listed ageing tests for bituminous binders thus:

- Extended Heating Procedures;
  - Thin Film Oven Test (TFOT),
  - Rolling Thin Film Oven Test (RTFOT),
  - Rotating Flask Test (DIN 52016),
  - Shell Microfilm Test,
  - Rolling Microfilm Oven Test (RMOT),
  - Tilt-oven Durability Test,
  - Thin Film Accelerated Ageing Test,
  - Modified Rolling Thin Film Oven Test,

- Oxidative (Air Blowing) Procedures;
  - Iowa Durability Test,
  - Pressure oxidation Bomb,
  - Accelerated Ageing test Device/Rotating Cylinder Ageing Test (RCAT),
  - Pressure Ageing Vessel (PAV),
  - High Pressure Ageing Test (HiPAT),

- Ultraviolet and Infrared Light Treatments;

- Microwave Ageing;

- Steric Hardening (no tests that address this phenomenon at the moment).

While ageing tests for asphalt mixtures could be:

- extended heating procedures,
- oxidation tests,
- ultraviolet/infrared treatment and,

- steric hardening.

Airey (2003) observed that the TFOT and RTFOT are the most commonly used short term ageing tests for simulating the hardening that occurs during asphalt mixture production. Currently, the most commonly used binder tests to simulate long-term ageing are the PAV and RCAT. In terms of long term ageing, no one test seems to be satisfactory for all cases and the RCAT method, based on a kinematic approach to ageing is probably the most acceptable. On the same note, Airey (2003) recorded that the most promising methods for short term ageing of asphalt mixtures are extended heating of the loose material and extended mixing and that the most promising methods for long term ageing of mixtures include extended oven ageing, such as the SHRP long-term oven ageing method, pressure oxidation, using low pressure oxidation as well as pressurised procedures, and ultraviolet and infrared light treatments. Meanwhile tests on mixtures could be conducted on either a compacted sample or a loose sample.

Bell (1989) on a comparative note, observed that research mostly has been focusing on bitumen, and that there has been little research on the ageing of asphalt mixtures, and to date, there is no standard test. Airey (2003) reasoned in line with this assertion when he stated that extended heating procedures show the most promise for short term ageing, and pressure oxidation and/or extended heating the most promise for long term ageing. He further advised that the condition for choosing whatever method used is that it should simulate conditions in the field and emphasize oxidation. He also remarked that the forced draft oven promises to be useful in extended oven ageing over the conventional ovens and however cautioned that, for mixtures, an elevated temperature level used in tests could cause specimen disruption. He also expressed a concern/need for the evaluation of the integrity of the compacted samples at high temperatures since it has been purportedly envisaged that the internal structure of a sample may be disrupted at temperatures such as 140°C and that in extreme cases the samples may slump. It must be noted that all these works have been carried out to simulate what obtains in the temperate regions of the world.

Oke et al. (2010) in an extensive study conducted with focus on hot tropical belt found out that as the binder used in their study (straight run paving grade binder 40/60 of Venezuelan origin with a penetration of 53dmm) aged, complex modulus, asphaltene content and softening point values increased while penetration values reduced irrespective of the ageing protocol i.e. either as bulk binder ageing or mixture ageing, in line with other researchers' findings. They observed that Ageing bituminous materials at 85°C over 120hrs (5days) in the forced draft oven did not yield the severity of ageing (below 6dmm) required to study heavily aged pavements typical of the tropics. This standard protocol yielded a residual binder of 23dmm penetration for the original 53dmm penetration binder used. However, it was able to age to 11dmm penetration after 840hrs (35days). It was observed that ageing bituminous mixture at a higher temperature of 125°C in the forced draft oven accelerates ageing, such that it only took 48hrs to reach a penetration of 11dmm compared to 840hrs (35days) for the standard protocol, thus saving laboratory time. Similarly they found out that ageing loose bituminous mixtures in the forced draft oven accelerates ageing. For example, ageing such mixtures at the standard temperature of 85°C over 24hrs yielded a pen of 31dmm in this work. Nguyen (2009) in his work in which he used compacted specimens with similar binder reported a similar penetration level of 31dmm after 120hrs of ageing. More importantly Oke et al. (2010) reported that the rheological properties and temperature susceptibilities of less severely aged binders (20dmm penetration and above), seemed to differ slightly. This was most obvious between the PAV aged binders and the recovered binders of mixtures aged at 125°C, and the properties of more severely aged binders are independent of the ageing protocol.

In conclusion, this review showed that temperature is an important factor for simulating accelerated ageing of asphalt mixtures in the laboratory. Also physical hardening of residual binder in the mix as a consequence of ageing could affect both the chemistry and rheology of bituminous binders. The review also showed that extended ageing of either compacted or loose asphalt mixtures in the forced draft oven are considered suitable to simulate both short and long term ageing in the laboratory. Ageing of compacted specimens is considered appropriate for cases where samples are to undergo further testing for mechanical properties such as stiffness modulus. Ageing of loose samples in particular is deemed appropriate for cases where aged materials are to be used as RAP in reconstituted asphalt mix for studies on recycled asphalt mixtures. Although ageing of compacted specimens is not used in the study discussed in this thesis, the softening point of the binder in the asphalt mix can be used as a guide for choosing ageing temperature to forestall specimen disruption. Ageing loose mixtures is a more practical means for achieving uniformly aged asphalt mix in a short laboratory time. This method also enables easy reconstitution of aged asphalt into new asphalt mix when compared to aged compacted specimens. It has the advantage that ageing can be conducted at high temperature, though there is the concern that excessive and high ageing temperatures can potentially damage the binder in the mix such that the physico-chemical and rheological properties of the binder are significantly different from what is obtained on the road pavement.

6   Cold Mixes

Generally, there are basically three types of asphalt mixes used in road construction. These are the Hot Mix (which is the conventional), the Cold Mix, and the Warm Mix. This piece focuses on cold mixes. The term cold is used in the sense that virtually all the operations involved in the production of the material are carried out at ambient temperature. Needham (1996) stated that cold mix is manufactured at ambient temperatures, although some processes can use the emulsion warmed to around 60°C. Meanwhile, the materials being used for the preparation of cold mixes are very similar to those in hot mix, and the major difference is that the bituminous binders used in cold mixes are liquefied and applied at relatively low temperatures compared to that of hot mix. Achieving these would mean that the binder is either emulsified or foamed. Since the use of water is normally involved in these two processes, hydraulic binders too are usually applied to facilitate among other things, the rapid evaporation of water in the mix. Thanaya (2003) listed the most commonly used types of Cold Bituminous Mixtures as:

- cold lay Macadam (cutbacks),
- grave emulsions (developed in France),
- foamed bituminous mixtures,
- CBEMs (Emulsified Asphalt Materials – USA) (Ibrahim and Thom, 1997).

Many researchers in this field are of the opinion that, in terms of energy savings, cold mixtures appear to be much more efficient than hot mixtures (Needham, 1996; Ibrahim, 1998; Thanaya, 2003; and Thom, 2008). Needham (1996) stated that, cold mix can be manufactured to cater for a range of different applicational regimes and that they are used mainly for base course and sometimes for binder or wearing course. They are applied through a number of methods ranging from hand application, through graders, finishers or pavers to self contained mixing and laying plants. Compaction regimes for cold mix are quite varied at present as different companies advocate and utilize different techniques. However, the preferred method seems to be steel rolling, followed by very heavy pneumatic tyred roller and finally finishing with steel. Meanwhile, PIARC (2003) stated that since cold mixes are difficult to compact, a continuous grading is suggested and that gap gradations and stone matrix are generally unsuitable for cold recycling. The grading envelopes used for the design of grave emulsion can be taken as a reference when using emulsion as the recycling agent in cold recycling.

In cold mixes, and particularly in the case of recycling, in addition to the specific area, it is necessary to take into account the capability of absorption of water and the chemical nature of the mineral surface. Needham (1996) opined that cold mix wearing course can be used on all but the most highly trafficked roads. Mixtures can be produced in specialised mixing plant, motor pavers or simple concrete mixers in which case the mixture can be produced on site. Also in line with Needham's assertions, but now considering the use of RAP, Carswell et al (2008) in an overview of the design guide and specification that is now the common practice for recycling in the UK, reported that cold recycling using a plant mixed process (*ex situ*) allows for screening and crushing of aggregates, prior to mixing with binder(s) in plant located nearby and the laying of materials in one or more layers using a paver. It also gives room for the introduction of alternative aggregates from sources other than the existing pavement. They stated further that, provided that the cold recycled materials can achieve the desired performance, the potential use of cold recycling is not limited. However, each site needs to be evaluated for the most appropriate maintenance selected in terms of:

- location,
- proximity of suitable location for setting up ex situ plant,
- proximity of source(s) of alternative materials, if required,
- type(s) and severity of deterioration,
- extent of deterioration,
- location of services within the pavement construction,
- condition of drainage,
- edge detail and verge condition.

Despite of all these advantages, cold mixes are still generally classified as inferior to hot mixtures with respect to performance, although the engineering equivalence and practical difficulties in adopting cold mixtures formulations have not yet been clearly defined (Biczysko, 1996). Carswell (2004) remarked that, looking at cold mixes, for example, despite the advantages for environment and health, they are still perceived as asphalt 2nd Class, used for reinforcement, re-profiling or for micro-surfacing, only where unavoidable, and only on secondary roads. Leech (1994) reported that, cold bituminous emulsions mixes are not so common in the UK because of the cold climatic conditions, more over in the UK there are sufficient HMA plants available and less remote areas- a serious challenge for CBEMs (Ibrahim and Thom, 1997; Khalid and Eta, 1997).

This position must have been premised on the fact that such a cold climatic condition would normally inhibit the rapid evaporation of water in the mix. Thanaya (2003) maintained that, cold mixes did not receive attention until 1992 in the UK. To this end, Zoorob and Thanaya (2002) remarked that Cold bituminous emulsion mixtures (CBEMs) are more universally accepted for low to medium traffic conditions, for works in remote areas and for small scale jobs such as reinstatement work. These notwithstanding, cold mixes have been used successfully in France, South Africa, Sweden and USA among other countries since the 1970s to meet various needs on the road (Thanaya, 2003; and Needham, 1996) and thus cold mixes hold a lot of promise. Carswell (2004) in buttressing this opined that recent experiences with cold mixes have been showing that they are far beyond the state of being assumed as asphalt 2nd class. He further stated that, an understanding of these mixes is improving, and it is sure that they are finding their ways to the high class, and thus would help gain a higher market share all over Europe. Zoorob and Thanaya (2002) in their work further strengthened this opinion in which they found out that, CBEMs with added cement at full curing can be comparable to conventional hot bituminous mixtures in terms of indirect tensile stiffness modulus.

In retrospect, it is worth noting that most of the previous works have been considering only virgin materials for cold mixes with the exception of Zoorob and Thanaya (2002) and Thanaya (2003) that looked at the incorporation of waste materials such as PFA, Red Porphyry Sand, Synthetic Aggregates, Steel Slag and Crumb Rubber into CBEMs. Most researchers only mentioned the possibility of using RAP in passing (Needham, 1996; Ibrahim, 1998; and Thanaya, 2003), and where they are used they are just regarded as black rocks without any regard to the properties of the residual binder in the RAP. The only related work to the author's knowledge is still ongoing at the University of Iowa, USA, which concerns cold-in place recycling typical of a temperate region (Lee, 2007). This book focuses on cold bituminous emulsion mixtures with RAP and virgin materials as aggregates in a circumstance surrounded by tropical condition and as well *ex situ* recycling.

7   Design of Cold Mixes

The importance of adequate design to all engineering structures cannot be overestimated. Without any prejudice, it is a commonly acknowledged fact that much advancement has been made in the design of hot bituminous mixtures compared to cold bituminous mixtures. While definite universal guidelines/standards have been established and are still being improved for the design of hot bituminous mixtures, Thanaya, (2007), Thanaya (2003),

Zoorob and Thanaya (2002), Ibrahim (1998), Asphalt Institute and Asphalt Emulsion Manufacturers Association (1997), FHWA (1997) and Needham (1996) among other researchers stated that there is no widely accepted mixture design method or structural design methodology for either virgin emulsion aggregate mixtures or cold recycled materials. This is one of the reasons why most pavement engineers feel more able to specify hot bituminous mixtures than cold bituminous mixtures.

This notwithstanding, FHWA (1997) affirmed that guidelines have been developed by several agencies, based on laboratory tests, empirical formulae or past experience with identical projects. It further stated that generally, the RAP particles are treated like black rock or aggregates in cold-recycled mix design and that however, the most commonly used recycling agent for complete cold recycling processes are emulsified asphalt cements (bitumen). This is because the emulsions are liquid at ambient temperatures and have the capacity for being dispersed throughout the mix and do not cause major air pollution problems, though quite a number of researchers have also found foamed bitumen useful in cold recycling (Jitareekul, 2009; Sunarjono, 2008; Jitareekul et al, 2007; Kim et al, 2007; Kim and Lee, 2006; Loizos and Papavasiliou, 2006; Long and Theyse, 2004; Romanoschi et al, 2003; PIARC, 2003; Jenkins, 2000; and GEOPAVE, 1993). Needham (1996), Asphalt Institute and Asphalt Emulsion Manufacturers Association, (1997), and Montepara and Giuliani (2002) remarked that the Marshall or Hveem design methods or modified versions are nearly always used on such occasions. Montepara and Giuliani (2002) opined that these design methods essentially investigate the best dosing of the bituminous emulsion and the total content of liquids (optimum fluid content), and that however, to study the same mixtures with added cement, it is necessary to consider other parameters, such as the water/cement ratio and the monitoring of the time to obtain certain mechanical properties. Needham (1996) opined that, if laboratory tests are not possible, empirical formulae for the addition level of emulsion to densely graded mixtures (see Table 5 for typical aggregates for dense-graded emulsion mixtures) are provided as a substitute and that the formula for the initial residual binder should be as shown below in equation 1:

$$P = (0.05A + 0.1B + 0.5\,C) \times (0.7) \tag{1}$$

Where:

P = Percent by weight of Initial Residual bitumen content by mass of total mixture

A = Percent of mineral aggregate > 2.36mm

Table 5: Aggregates for Dense-Graded Emulsion Mixtures (Asphalt Institute and Asphalt Emulsion Manufacturers Association, 1997)

Sieve Size	Semi-Processed Crusher, Pit or Bank Run	Processed Dense-Graded Asphalt Mixtures Percent Passing by Weight				
50.0mm (2 in.)	-	100	-	-	-	-
37.5mm (1-1/2 in.)	100	90-100	100	-	-	-
25.0mm (1 in.)	80-90	-	90-100	100	-	-
19.0mm (3/4 in.)	-	60-80	-	90-100	100	-
12.5mm (1/2 in.)	-	-	60-80	-	90-100	100
9.5mm (3/8 in.)	-	-	-	60-80	-	90-100
4.75mm (No. 4)	25-85	20-55	25-60	35-65	45-70	60-80
2.36mm (No. 8)	-	10-40	15-45	20-50	25-55	35-65
1.18mm (No. 16)	-	-	-	-	-	-
600μm (No. 30)	-	-	-	-	-	-
300μm (No. 50)	-	2-16	3-18	3-20	5-20	6-25
150μm (No. 100)	-	-	-	-	-	-
75μm (No. 200)	3-15	0-5	1-7	2-8	2-9	2-10
Sand Equivalent, Percent	30 min.	35 min.	35 min.	35 min.	35 min.	35 min.
Los Angeles Abrasion @ 500 Revolutions	-	40 max	40 max	40 max	40 max	40 max
Percent Crushed Faces	-	65 min.	65 min.	65 min.	65 min.	65 min.

B = Percent of mineral aggregate <2.36 and >0.75mm

C = Percent of mineral aggregate < 0.75mm

Thanaya (2007) having reviewed and experimented with some guidelines for the design of cold mix in his work proposed that cold mix design should consider:

1 Determination of aggregate gradation, which should be continuous due to its interlock property, giving strength to the material during early life strength of cold mixes. This can be obtained using Cooper's equation:

$$P = \frac{(100-F)(d^n - 0.075^n)}{D^n - 0.075^n} + F \qquad (2)$$

Where P is the percentage material passing sieve size d (mm), D is maximum aggregate size (mm), F is the percentage filler, and n, an exponential value that dictates the concavity of the gradation line.

2 Estimation of initial residual bitumen content and initial emulsion content in line with equation (3.1) above.

3 Coating test or binder compatibility test.

4 Determination of compaction level to meet porosity target;
   a. Storage time for the loose mixture and condition- sealed and unsealed,
   b. Compaction by applying an initially judged compaction effort
   c. Curing for dry density determination,
   d. Determination of specific gravity of the mix and porosity value after obtaining dry density data,
   e. Adjustment of compaction effort and determination of compaction effort to meet the porosity target. Suggested porosity target is 5-10%. The following useful formulae were also suggested:

$$SG_{mix} = \frac{100}{\frac{\%CA}{SGCA} + \frac{\%FA}{SGFA} + \frac{\%F}{SGF} + \frac{\%Binder}{SGBinder}} \qquad (3)$$

Where:

SG_{mix} = Specific Gravity of the mix

%CA = Percentage of Coarse Aggregates

%FA = Percentage of Fine Aggregates

%F = Percentage of Filler

SGCA = Specific Gravity of Coarse Aggregates

SGFA = Specific Gravity of Fine Aggregates

SGF = Specific Gravity of Filler

By weight of total mix:

$$Bulk\ Density = \frac{\%Weight\ in\ air}{WeightSSD - Weight\ in\ water} \qquad (4)$$

Weight saturated surface dry (SSD) is obtained by towel drying the samples after weighing in water, until no air bubble occurs.

$$\text{Porosity}(P - \%) = \left(1 - \frac{\text{Bulk Density}}{\text{SG}_{\text{mix}}}\right) \times 100\% \tag{5}$$

5 Variation of residual bitumen content.
6 Determination of optimum residual bitumen content.
7 Calculation of bitumen film thickness at optimum residual bitumen content. The bitumen film thickness (BFT) can be calculated using the formula:

$$\text{BFT} = \frac{\%\text{Binder}}{100 - \%\text{Binder}} \times \frac{1}{\text{SGBinder}} \times \frac{\%\text{F}}{\text{ASA}} \tag{6}$$

Where ASA is the aggregate surface area and requires surface area factor as given Table 6, the calculation of the aggregate surface area (ASA) is obtained as shown Table 7. The ASA is calculated by multiplying the total percent passing each sieve size by the appropriate SAF and adding together. The minimum BFT to be targeted is 8 micron.

8 Determination of retained stability of the mixture at optimum residual bitumen content only, according to the design curing procedure.
9 Determination of the ultimate strength at full curing of the samples at optimum residual bitumen content only.
10 Incorporation of about 1-2% cement by mass of aggregates to improve the performance of the cold mix.

Also, a general flow chart for the design of mixtures as suggested by Jenkins (2000) is shown in Figure 3.

Although Thanaya (2007) concluded that this design method was found to be simpler than those reviewed, his work was conducted in the light of the conditions

Table 6: Surface Area Factor (SAF) (Thanaya, 2007)

Particle/Sieve Size	Surface Area Factor (m²/kg)
Maximum size (all sizes greater than 4.75mm)	0.41
4.75mm (No. 4)	0.41
2.36mm (No. 8)	0.82
1.18mm (No. 16)	1.64
600µm (No. 30)	2.87
300µm (No. 50)	6.14
150µm (No. 100)	12.29
75µm (No. 200)	32.77

Table 7: Calculation of Aggregate Surface Area (ASA) (Thanaya, 2007)

Sieve		ASA Calculation		
Inch/No.	mm*	Estimated Total Pass (%)**	SAF	ASA (m^2/kg)
		a	b	c = a x b
3/4″	19.0	100	0.41	0.4100
3/8″	9.5	-		
No. 4	4.75	58	0.41	0.2378
No. 8	2.36	41.5	0.82	0.3403
No. 16	1.18	28.8	1.64	0.4723
No. 30	600µm	19.6	2.87	0.5625
No. 50	300µm	12.7	6.14	0.7798
No. 100	150µm	7.7	12.29	0.9463
No. 200	75µm	4	32.77	1.3108
ASA (sum)				**5.0598**

* In line with particle size/sieve in Table 7
**Estimated based on the mixture's aggregate grading curve from Cooper's formula curve
i.e. equation 3.1

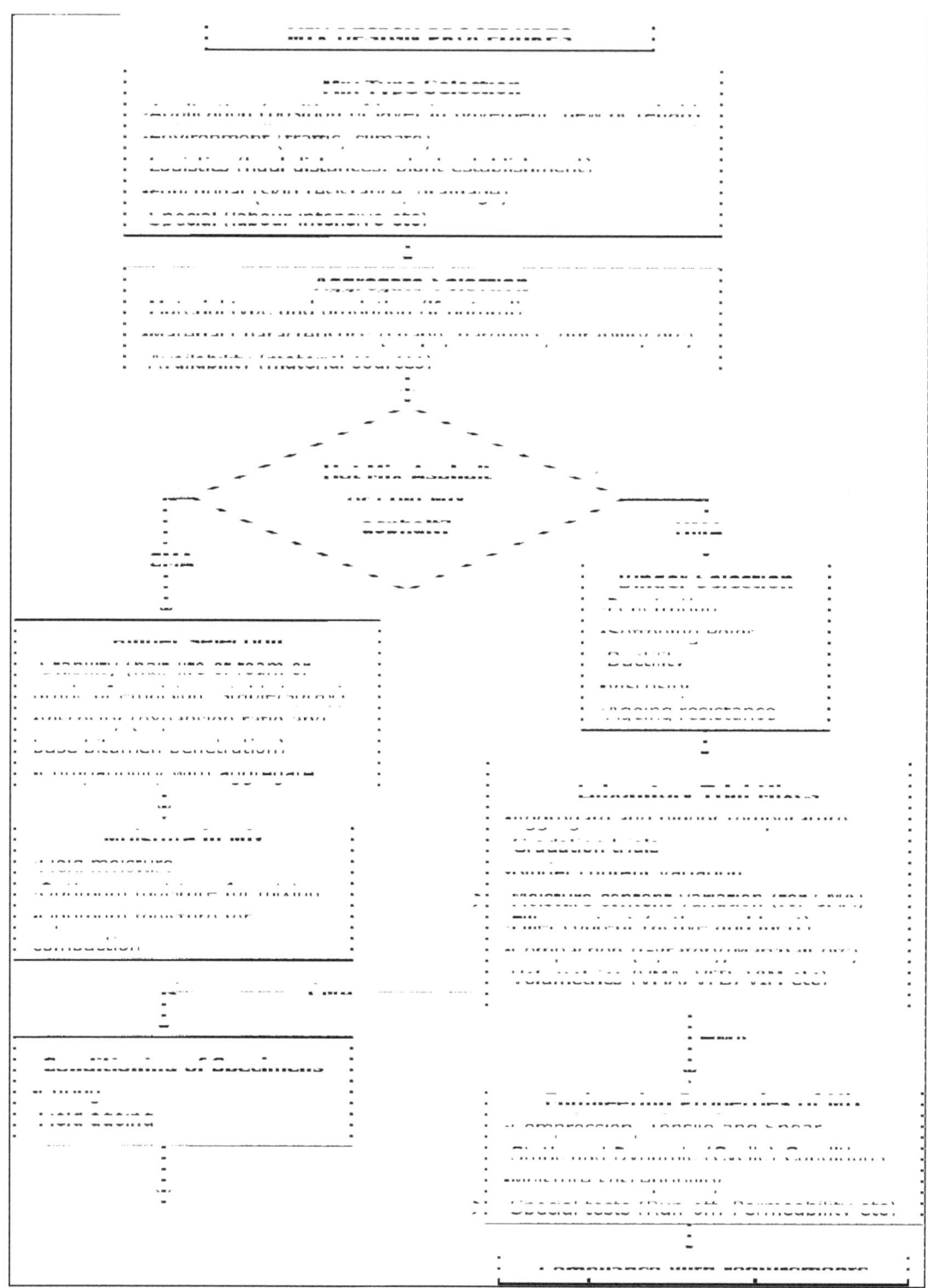

Figure 3: A General Procedure for Mix Design (Jenkins, 2000)

26

that obtain in temperate regions. In the same vein, while he used cationic bitumen emulsions along with limestone, RAP and granite were never experimented with in his work.

8   Curing Protocols

Unlike hot mixes, adequate and proper curing is a requirement for cold mixes (Thom, 2009). Jenkins (2000) defined curing of cold bituminous mixes, whether emulsion or foam as the process whereby the mixed and compacted material discharges water through evaporation, particle charge repulsion or pore-pressure induced flow paths. A reduction in moisture content leads to an increase in strength of the mix (both tensile and compressive). Roberts et al (1984) found a significant increase in tensile strength as curing temperature was increased from 23ºC to 60ºC. Bocci et al (2002) discovered that as curing time and curing temperature increase, the performances of the mixtures considerably improve; curing for 14 days at 20°C is almost equivalent to 7 days at 40°C. Thanaya, (2007), Leech (1994) and Santucci (1977) reported that full curing can occur between 2-24mths in the field. Table 8 details some curing methods while Figure 4 shows the effect of different curing conditions and compaction on the compressive strength of reference grave emulsion.

Jenkins (2000) consequently opined that the temperature of curing cannot be ruled out as unimportant to mix preparation as temperature and moisture are dependent variables with temperature influencing the rate of moisture loss. Meanwhile, Jenkins (2000) had earlier submitted that too little moisture impedes dispersion of the foam (emulsion), workability and compaction of the mix and too much moisture increases the curing time and reduces the density and strength of the compacted mix and that the moisture contents of mixes that are oven cured in an unsealed state are generally between 0% and 1.5% and always less than 4%, which is seldom representative of field conditions. In addition, the influences of curing temperature on changes in the binder condition have not been analysed in the literature, which is unfortunate considering the high surface area of the binder and higher void contents in foamed mixes. The challenge is to select the appropriate curing conditions in the laboratory, ensuring adequate shear strength in early life and selecting the correct stiffness for the structural design life. In addition, compaction due to traffic requires consideration. In a related view, Serfass et al (2003) believed that evaluating cured cold mixes in the laboratory is clearly necessary, but reproducing exact field curing conditions is too complicated and, above all, time consuming, and thus an accelerated curing method is

Table 8: Some Curing Methods for Foamed Bitumen (Jenkins, 2000)

Curing Method	Equivalent Field Cure	Reference
3 days @ 60°C + 3 days @ 24°C	Unspecified	Bowering (1970)
3 days @ 60°C	Construction period + early field life	Bowering and Martin (1970)
3 days @ 60°C	Between 23 and 200 days from Vane Shear Tests	Acott (1980)
1 day in mould	Short term	Ruckel et al (1982)
1 day in mould + 3 days @ 40°C	Between 7 and 14 days (intermediate)	Ruckel et al (1982)
1 day in mould + 3 days @ 40°C	30 days (Long term)	Ruckel et al (1982)
1 day @ 38°C	7 days	Asphalt Institute
10 days in air + 50 hours @ 60°C	Unspecified	Van Wijk and Wood (1983)
3 days @ ambient temp. + 4 days vacuum desiccation	Unspecified	Little el al (1983)
3 days @ 23°C	Unspecified	Roberts et al (1984)
3 days @ 60°C	Unspecified	Lancaster et al (1994)
3 days @ 60°C	1 year	Maccarrone et al (1994)

Note: 1. Specimens are cured in an unsealed state in the oven, unless otherwise stated.

2. Brennen et al (1981) developed the procedure to first cure the foamed specimens in the mould for 24hrs during the most fragile period.

3. Vacuum desiccation methods are in line with the Asphalt Institute design procedure (PCD-1) and require further investigation.

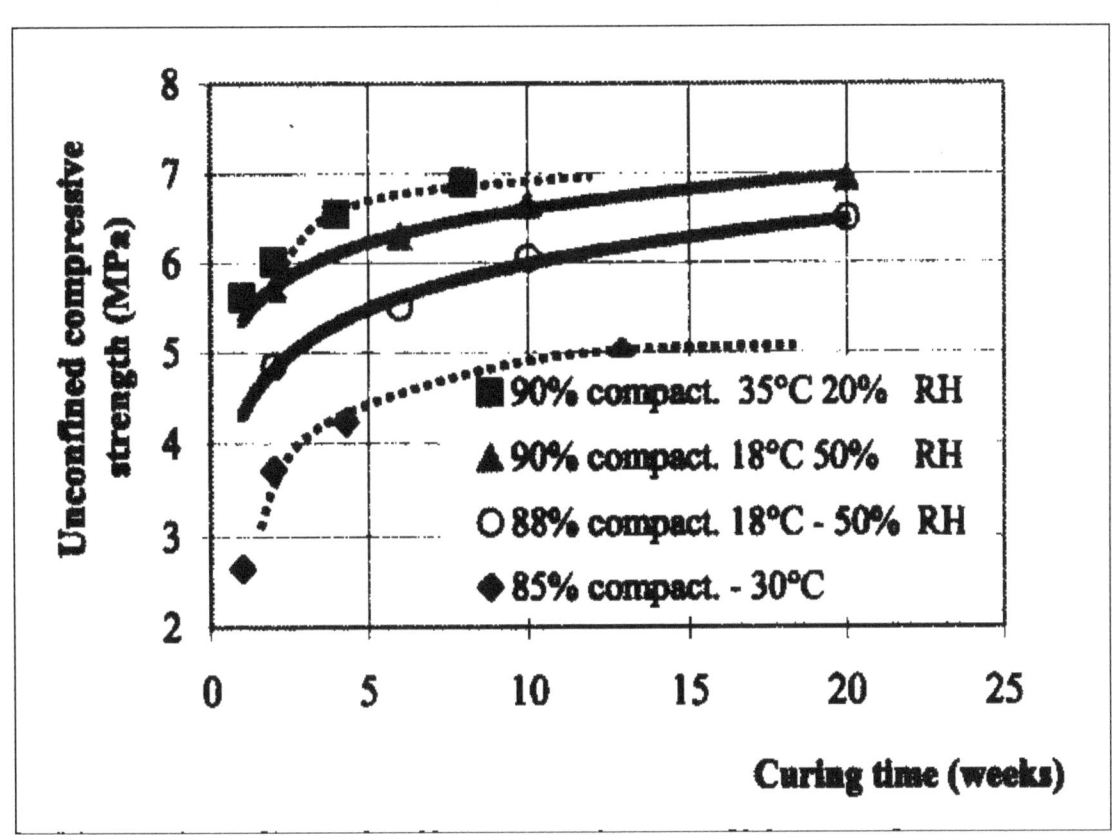

Figure 4: Effect of Different Curing Conditions and Compaction on Compressive Strength of Reference Grave Emulsion (Serfass et al, 2003)

necessary. The requirements are that:

- The curing procedure should be as short as possible,
- It must produce materials in a state as close as possible to their in-place mature state,
- It must not cause significant ageing of the bituminous binder,
- The laboratory equipment should not be too sophisticated.

Serfass et al (2003) further discovered in their study that the moisture content of small specimens decreases very quickly, whatever the temperature. For large specimens it takes longer. Also, too high a temperature (e.g. 50°C) tends to produce too quick a sample drying, which can cause cracking of large samples. They advised that, it should be borne in mind that, in the field, cold mixes rarely become totally dry. The moisture content has often been found to be between 0.5 and 1.5% in road bases in temperate climates. They established a relationship between the degree of compaction and stiffness modulus and thus stated that modulus values must therefore always be related to the degree of compaction of the mix tested. Therefore in order to obtain cold mixtures in a mature state, they proposed that such samples should be cured for 14 days at 35°C-20% Relative Humidity (RH). This procedure does not cause deterioration to the specimens. 35°C was chosen because it is realistic, as it prevails for long periods in the actual pavements. It was further proposed that curing over 14 days at 18°C – 50% RH would be appropriate for mix in a fresh state i.e. a few weeks after laying (18°C chosen because that is a typical average daily temperature for temperate climates).

9   Some Compaction Methods for Cold Mixes

Compaction is another important factor that is pivotal to the efficient performance of cold mixtures. In buttressing this, Harun and Morosiuk (1995) remarked that the full benefit from the use of bituminous materials in road construction can only be achieved if a satisfactory degree of compaction is attained during construction. They further opined that if shear failure is to be prevented then secondary compaction of the material after construction must be limited to a safe value and that an alternative to the use of additives is a modification of the mix design and compaction method to produce a more mechanically stable material, again making it more resistant to deformation at high temperatures. They thus proposed that design of mixes should be done using refusal density.

While Thanaya (2007) suggested that compaction should be done to target a porosity within the range 5-10% generally for cold mixes which thus implies heavy compaction effort, Jenkins (2000) opined that the distribution of binder within a foamed mix differs from that of HMA and the inclusion of the water phase sets these two mixes apart, in so doing introducing differences in compactability, and that a laboratory compaction technique that not only achieves the void content expected in the field, but also emulates the particle orientation after rolling is to be sought. He further advised that both the volumetrics and the engineering properties of the mix require consideration in the selection of an appropriate compaction technique and more reliable links between laboratory and field compaction are required. PIARC (2003) advised that, for the laboratory study of mix design for recycling, the total fluid content to insure adequate compaction of the mix must be determined. AEMA and AI (1997) remarked that for gradations containing appreciable fines, aeration or drying prior to compaction may be required. FHWA (1997) suggested that the rate at which the reaction between the recycling agent and the aged asphalt occurs is a function of the properties of the recycling agent and the aged asphalt cement, and the mechanical effects of the physical processes such as mixing, compaction, traffic and climatic conditions.

Thanaya (2007) further opined that the porosity of CBEMs can be reduced to meet a pre-selected target simply by increasing the compaction effort and that the compaction effort is a significant variable that needs to be determined depending on the target porosity, mixture type, storage conditions and storage time prior to compaction. Similarly, Kim et al (2007) categorically listed compaction method as one of the factors to be considered in design.

All these assertions of previous researchers indicate that compaction method employed in the laboratory is very important in the design of mixtures. However, the situation becomes more confusing as Needham (1996) stated that the field compaction regimes for cold mix are quite varied at present as different companies advocate and utilize different techniques and that however, the preferred method seems to be steel rolling, followed by very heavy pneumatic tyred roller and finally finishing with steel. To ascertain which compaction method is the best, researchers have made every effort to produce compaction methods in the laboratory that simulate what obtains in the field. Meanwhile, a lot of methods for the compaction of such mixtures in the laboratory have been developed. Needham (1996) listed them as the Marshall Hammer method, The Percentage Refusal Density Apparatus, the Static Load Press, the Roller Compactor and the Gyratory Compactor. Ibrahim (1998) also used the Vibrating Compactor in his work but stated that Roller Compactors have not been used in the study of cold mixes. Some of these methods are detailed in Table 9.

Obviously, not all these methods have been found to be successful/applicable in all conditions and thus efforts have been made to identify the most suitable method which among other things mimics the compaction on the road, simple to operate and also affordable for different operations and as well materials. In the light of this, various studies have been conducted to ascertain best methods for compacting mixtures and as well to understand the effect of compaction on mixtures. Kim et al (2007) discovered from their studies that additional foamed asphalt helped mixtures compact better under gyratory compaction, but not under Marshall Hammer compaction. Smith and Jones (1998) in an earlier work reported that the 75-blow Marshall compaction underestimates the effect of secondary compaction under traffic and many of these surfacings suffer severe structural instability leading to plastic deformation. In line with this, Kim and Lee (2006) observed that 75 blows from the Marshall Hammer did not provide sufficient compaction to simulate initial compaction after construction. They opined that although the gyratory compaction method produced a higher density value than the Marshall Compaction method, the samples made by the gyratory compactor exhibited lower resilient modulus values than the ones prepared by the Marshall compactor.

In another similar work, Ulmgren (2003) observed that with an increased angle of compaction a predetermined degree of compaction is reached with fewer rotations. Figure 5 shows this relationship. On another note, Thanaya (2007a) stated that in order to achieve an air void content of between 5-10%, compaction was done using 240 gyros- extra heavy compaction (medium compaction is carried out at 80 gyros which is considered to be the equivalent of 50 blows using the Marshall hammer, while heavy compaction is carried out at 120 gyros, equivalent to 75 blows using the Marshall hammer). He concluded that the application of a heavier compaction level is inevitable in cold mixes, as the emulsion set and hence the mixes stiffen during compaction. Serfass et al (2003) in their work established a relationship between the degree of compaction and stiffness modulus. Figure 6 shows this relationship. Thanaya (2003) in his work observed that, compaction of CBEMs in two layers aided escape of moisture faster which eventually reduced the curing time.

Recently, Lee and Kim (2006) examined the compaction characteristics of RAP

Table 9: Summary of Laboratory Compaction Techniques used for Foamed Mix Design (Jenkins, 2000)

Compaction Method	Settings/Temperature	Remarks	Reference
Kneading Compactor	Ambient temperature	-	Shackel (1974)
Kneading Compactor	Ambient temperature	-	Bowering & Martin (1976)
Gyratory Compactor	Angle = 1° Ram pressure = 1.38 MPa	Optimum Bitumen Content = f(Degree of comp)	Tia and Wood (1982)
Texas Gyratory Compactor	25°C	-	Little et al (1983)
Gyratory	20 rev. with Ram pressure =1.38 MPa	12% higher density than 75 blows Marshall	Brennen et al (1983)
Gyratory Compactor	150 cycles, Angle = 2° Ram pressure = 0.24 MPa for 100mmΦ 150 cycles, Angle = 3° Ram pressure = 0.54 MPa for 150mmΦ	-	Maccarrone et al (1984)
PCG (French Gyratory Compactor)	200 cycles at French standard settings	LCPC carousel: PCG 200 gyrations ≡ 85% Solid density	Brosseaud et al (1997)

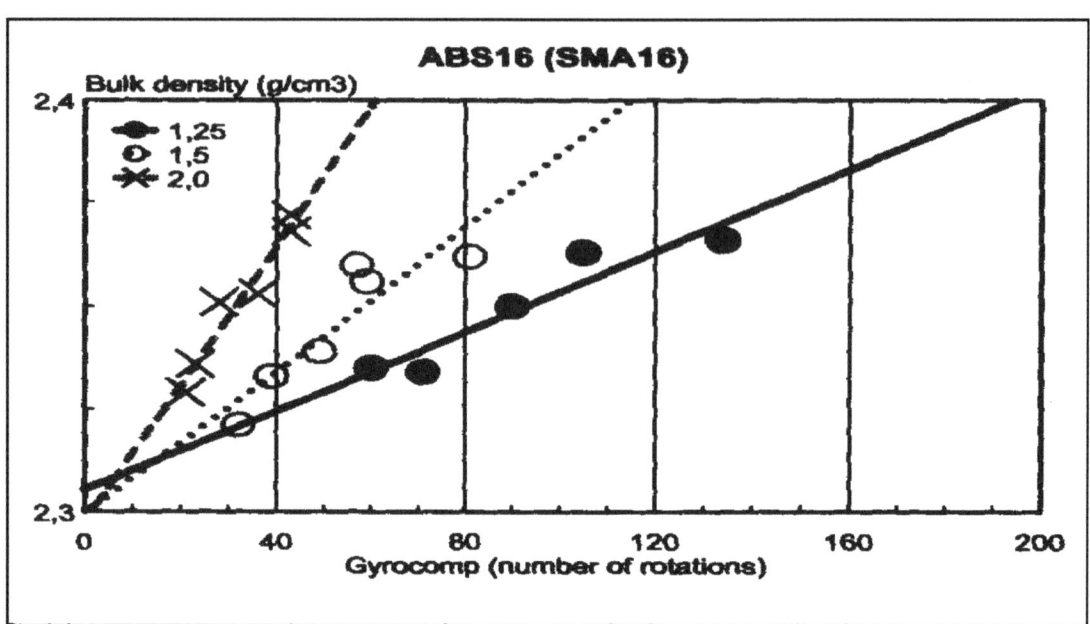

Figure 5: Relationship between Bulk Density and Number of Rotations (note, with an increased compaction angle i.e. 1.25 to 2°, the number of rotations needed to reach a given density is fewer, ABS is the Swedish equivalent of SMA) (Ulmgren, 2003)

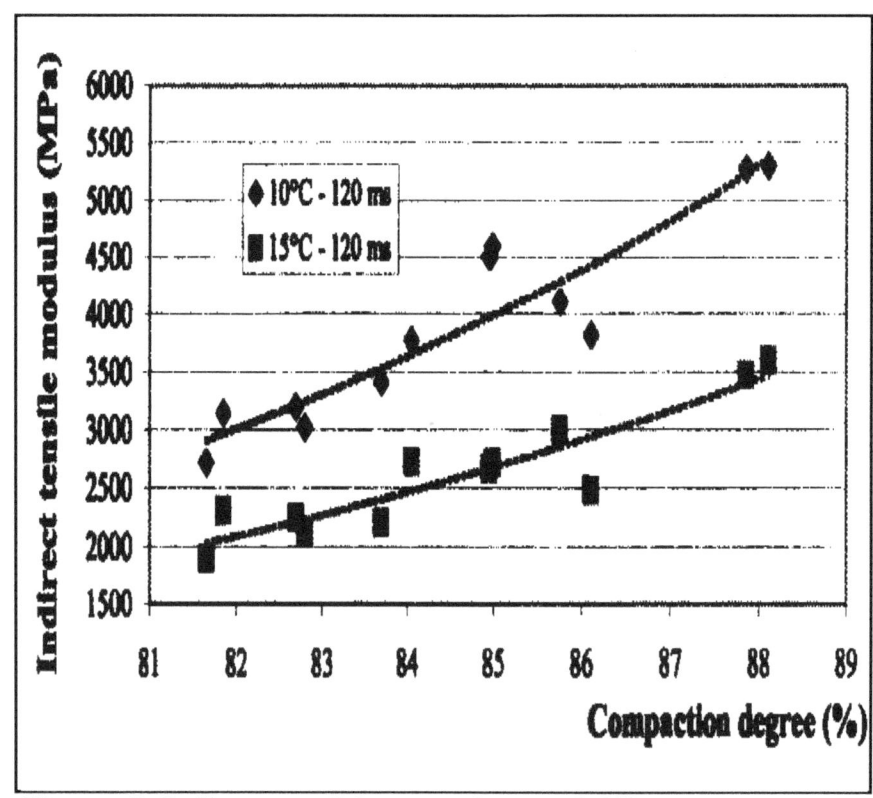

Figure 6: Influence of Compaction on Stiffness (Serfass et al, 2003)

material at different temperatures and moisture contents, RAP materials were compacted at temperatures of 25°C, 40°C and 55°C and moisture contents of zero and 4% using the gyratory compactor without adding any additional bitumen. The three temperatures were selected to represent the pavement surface temperatures in the field for early fall, late spring, and peak summer conditions respectively. Two samples were compacted up to 200 gyrations for each of six cases of three RAP temperatures and two moisture contents. The bulk specific gravities of compacted RAP samples with and without water were measured and plotted against the number of gyrations. The compacted samples with 4% water exhibited the higher bulk specific gravity than those without water and the higher RAP temperature also produced the higher bulk specific gravity.

Also, Lee and Kim (2007) investigated the compaction characteristics of RAP materials, and as a reference point, RAP materials were compacted using a gyratory compactor without adding water or foamed asphalt. They found a significant increase in bulk specific gravity by adding foamed asphalt. Moisture content was fixed at 4%. Specimens compacted by gyratory compactor at 30 gyrations or by Marshall Hammer at 75 blows were cured at 40°C for three days or 60°C for two days. The indirect tensile strength of gyratory compacted and vacuum saturated specimens was more sensitive to foamed asphalt contents than that of the Marshall hammer compacted vacuum saturated specimens. Brennan et al (2007) concluded in their work on compaction that, the voids content of a cold mixture can be up to 9% greater than that for a comparable hot mixture.

Indeed the development/adoption of compaction method in the laboratory which mimics what obtains in the field is essential for coming up with efficient designs most especially for cold mixes. As discussed earlier, it would be observed that a lot has been done on ascertaining the best method to adopt for simulating in the laboratory the compaction that takes place in the field. Most researchers have been favourably disposed to the gyratory compactor, which was developed under the SHRP programme in the USA. However, as most researchers rightly caution, both the volumetrics and the engineering properties of the mix require consideration in the selection of an appropriate compaction technique and more reliable links between laboratory and field compaction are required.

10   Performance Characteristics of Cold Mixes

Quite a number of investigations have been carried out to examine the performance of cold mixes. Through the outcomes of some of these investigations, the major problems of cold mixes have been identified, and procedures for mitigating them have been proposed. Meanwhile, performances of cold mixes are normally examined with reference to hot mixes. In a recent work, Thanaya (2007) observed that CBEMs compared well in mechanical properties with hot mixes as detailed in Table 10. On a similar note, Robinson (1997) using cores examined in the UK in his work as reported by Thanaya (2003), revealed that mixtures develop stiffness gradually, e.g. the ITSM value of a 6mm Dense Macadam (200pen base emulsion) met the specification requirements of 600MPa after 10mths and the value increased to almost 800MPa after 24mths. Thanaya (2007) further noted that considering the creep slopes of the cold mixes studied in the work, it was confirmed that cold mixes are suitable for low to medium trafficked roads and that the addition of 1-2% cement (see Tables 3.11a and b) by mass of aggregates into cold asphalt emulsion mixes significantly improves the overall mechanical properties of the CBEMs.

Brown and Needham (2000) in buttressing this submitted that adding OPC to bitumen emulsion mixtures (cold mix) has some beneficial effects. They revealed that without OPC, cold mix failed at less than 1000cycles in the unconfined mode of Repeated Load Axial Tests (RLAT), but suggested that it could do better if the triaxial mode is used. They further found out that OPC has no beneficial effect on the performance of hot mix and that cold mix with OPC offered better resistance to permanent deformation than the hot mix. Also, OPC acts as a secondary binder and the addition of OPC caused a reduction in fatigue life above 200 microstrains. They however, opined that a pavement structure might not in real life be subjected to as high a microstrain level, thus it could be said that OPC addition clearly extends the fatigue life of a pavement. Thanaya (2007) later confirmed this in his work (the developed relationship is as shown in Figure 7) though Ibrahim (1998) opined that there is no consensus for dealing with fatigue in cold mixes.

For example, Khalid (2003) observed that CBEMs continue to be underutilised in the UK in comparison with other European countries despite their significant environmental and economic potential. He opined that the main reasons behind this under-use are believed to be the high cost of emulsion binders and complexity involved in the design and performance assessment of CBEMs. More so, pavement engineers would always want to put in place structures that would be able to perform their intended design roles immediately after construction. Serfass et al (2003) submitted that cold mixes are evolutive materials (see Figure 8), especially in their early life and that their peculiar behaviour results from the

Table 10: Properties of the CBEMS Compared to Hot Mixes (Thanaya, 2007)

Mixture type	Compaction effort	Porosity (%)	ITSM (MPa)
CBEM	2 x heavy compaction	9.7	2275 (full curing)
CBEM + 1% cement*	2 x heavy compaction	9.4	3378(full curing)
CBEM + 2% cement*	2 x heavy compaction	9.2	4970(full curing)
100pen Hot Mix	Medium compaction	4.7	2150
100pen Hot Mix	Heavy compaction	3.4	2520
*by mass of aggregates			

Table 11a: The Beneficial Effect of Adding Cement to CBEMs (Thanaya, 2007)

No	Type of mix, with Nynas emulsion	ITSM after 1 month (MPa)	ITSM after 2 months (MPa)
1	WC 1-RPS without cement	752.37	816.04
2	WC 1-RPS + 2% OPC	1691.1	2084*
3	WC 1-RPS + 2% natural cement	1456.95	1691
4	WC 1-RPS + 2% rapid setting cement	1769.45	2258*
*Had achieved target			

Table 11b: Beneficial Effect of Adding Cement to CBEMs (Thanaya, 2007)

No	Type of mix	Porosity (%)	ITSM (MPa)
I	CM with Nynas emulsion		
I.1	CM without cement	9.2*	1595
I.2	CM + 2% OPC	8.7*	2581*
	CM + 2% rapid setting cement	8.5*	2593*
II	CM with TotalfinaElf emulsion		
II.1	CM without cement	8.2*	1346
II.2	CM + 2% OPC	7.8*	2327
	CM + 2% rapid setting cement	7.9*	2696
*Meets target			

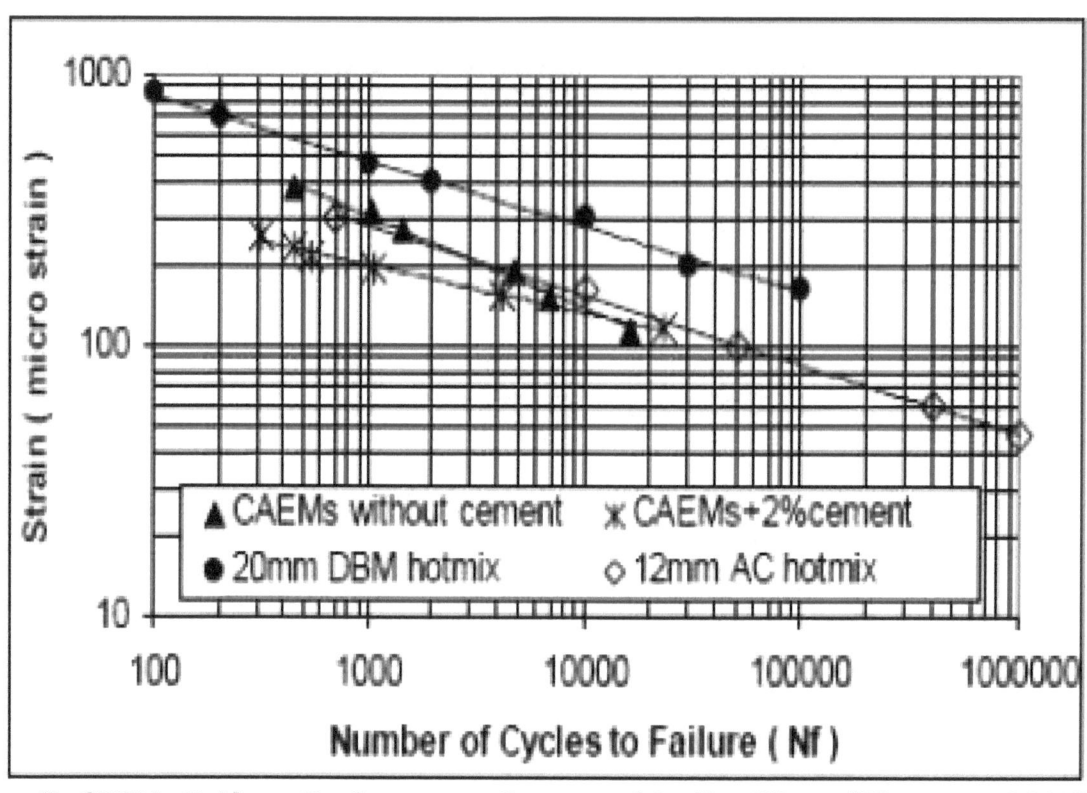

Figure 7: CBEMs Fatigue Performance Compared to Hot Mixes (Thanaya, 2007)

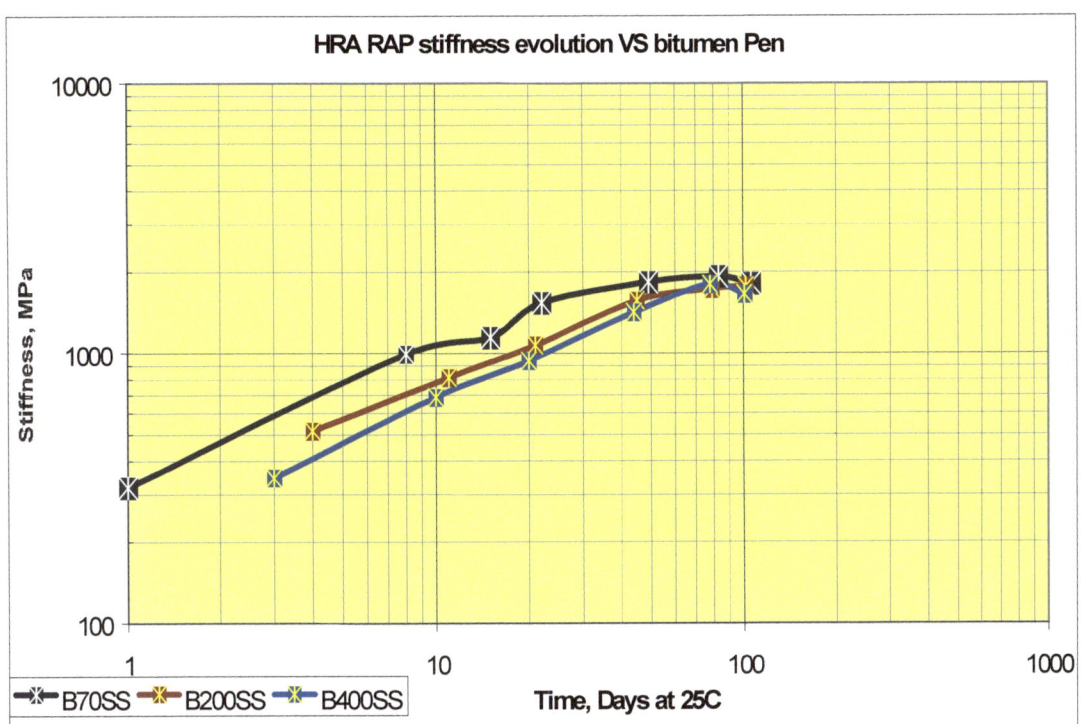

Figure 8: Influence of Bitumen Grade on Cold RAP Mix Stiffness versus Time Showing the Evolutive Behaviour of Cold Mixes (Lesueur et al, 2005; and Walter et al, 2008)

combination of several factors, presence of water, aggregate emulsion reactivity, binder film coalescence and cohesion build up. In the field, cold mixes reach their mature level of properties only after a period of time. In temperate climates and under medium traffic, at least one complete cycle of seasons is necessary for the mix to attain a stable condition. The curing time may be longer if the climate is cooler and more humid, or shorter if the traffic is lighter. This problem has been successfully addressed by applying hydraulic binders to the mix. However, when his option is used, care must be taken in the choice of hydraulic binder since they can significantly shorten the opening time for construction (depending on the type of cement used).

However, the major problems of cold mixes which would normally require the exercise of caution during the design and construction stages have been identified too. For example, Zoorob and Thanaya (2002) stated that besides being highly sensitive to ambient conditions during the mixture laying stage, the three most common problems encountered when using cold mixes for road layers are:

- High mixture porosity,
- Low early life strengths,
- Long curing times especially in cold and damp environments.

They identified other problems such as poor aggregate coating, binder stripping and low binder film thickness which have now been overcome to some extent with rapid improvements in emulsion manufacturing technology and site production techniques.

Thanaya (2003) established that this binder coating problem is particularly peculiar to the coarser aggregates and that the performance of the mixtures is highly dependent on the availability of a superior emulsion formulation. Jove and Bock (2003) observed that the weak aspect of cold recycling is poor cohesion behaviour of the final mixture. Thanaya (2003) discovered that CBEMs can have some problems with binder drainage during storage as a result of low binder viscosity. Binder stripping can also occur due to weak adhesion and in general, the compacted mixtures contain high void contents.

On a similar note, Bocci et al (2002) established in their work that RAP particle size distribution has little influence on the compressive strength of cold recycled bituminous concrete. They also found out that the presence of lumps in RAP has a negative effect on the performance of the mixtures. Epps (1990) and Roberts et al (1984 & 1991) considered that cold mix recycling is only used to form a base course for low –to- medium traffic volume highways, because cold mixtures are not structurally as strong as hot mixtures and cold recycled mixes do not have adequate resistance to either abrasion by traffic or moisture induced damage.

Staple (1997) contrary to the experience of Robinson (Thanaya, 2003) was not favourably disposed to the use of emulsion due to the fact that cold emulsion mixes could not meet UK standards even after 18months of being laid as the stiffness moduli were far too low. He observed that cold emulsion macadam had significantly lower elastic modulus than that required. However nothing was stated about materials used and the factors that guided selection. He further opined in this work that the high air void content of the mix could have been responsible for the general low stiffness. Room is generally given for some degree of permeability and contact with air to allow for the curing or break of an emulsion. Staple (1997) was still doubtful about the applicability of cold mixes in road construction.

It may be observed from this review that there are a lot of challenges to overcome if cold mix, most especially cold recycled mix is to be fully integrated into road construction. Obviously, all of these observations have been made in the light of the performances of such mixes under temperate conditions. In simulating what obtains under a hot climate (temperature), one challenge is to determine which laboratory test would be the most appropriate for characterising the mechanical responses of recycled emulsion cold mixtures-since Ibrahim (1998) opined that such materials exhibit composite responses i.e. during the preparation of the mixes and laying, they behave as unbound granular materials, while after construction and having gained appreciable strength they start acting as bound materials. Other challenges in this regard are mentioned in the following section.

11   Interaction between Aged and Virgin Binders in Cold Recycled Mixes

While a lot has been done by researchers to establish the relationship between aged binders and recycling agents in hot recycled mixes, not much has been done on cold recycled mixes. Quite a number of researchers have established beyond reasonable doubt that in hot recycled mixtures, there is an interaction between aged binders and recycling agents (Lee et al, 1983; Noureldin and Wood, 1987; Soleymani et al, 2000; McDaniel et al, 2000; Karlsson and Isacsson, 2003; Schiavi et al, 2003; Romera et al, 2006; Artamendi and Khalid 2006; Chen et al, 2007; and Shen et al 2007).

On the other hand, FHWA (1997) stated that at ambient temperature, the softening effect of the recycling agent is a time and temperature dependent physico-chemical process and thus, the rate at which the reaction between the recycling agent and the aged asphalt cement (bitumen) occurs is a function of the properties of the recycling agent and the aged asphalt cement, and the mechanical effects of the physical processes such as mixing, compaction, traffic and climatic conditions. The relative contributions of the recycling agent and the aged asphalt binder are not fully understood at this time and thus, FHWA (1997) assumed that for cold recycled mixes, RAP only acts as black rock- which implies that the aged binder in the RAP does not interact with the recycling agent.

The only work among the reviewed reports which shed light on the interaction between aged binders in cold mixes and recycling agents was carried out very recently under the SCORE project in which Lesueur et al (2005) and Walter et al (2008), using the DSR (see Figure 9) and measurements of the mechanical properties of binders (aged and rejuvenating agents) and cold recycled mixtures respectively, established that ultimately there is an interaction between aged binders and the recycling agents. They studied the effect of new bitumen grade on complex modulus versus time and discovered that all samples seem to converge after 100days which indicate that, due largely to diffusion, the 400 and 200 Pen aged bitumen evolve toward a harder grade, between 70 and 100 penetration. However, the 70 penetration binder provided the highest modulus values both initially and for the next 100 days.

On the maturation time, they observed that cold diffusion is a slow process, the improvement between 1 week and 1 month was negligible which tends to indicate that most of the diffusion at ambient temperature takes place during the first 7 to 10 days. They however concluded that, DSR measurements are limited for practical reasons to temperatures above 40°C and reducing the number of concentration measurements is highly desirable given the large experimental times involved. They showed in their studies that it is possible to follow the diffusion phenomenon between new and old bitumen by measuring the evolution of stiffness of cold RAP mix versus time as the stiffness of a mix is directly related to the binder consistency and curing state of the emulsion binder. They also recommended the use of rejuvenating oil for RAP containing more than 4.5% residual bitumen and also hard grade bitumen emulsion when stiffness is the key performance criterion.

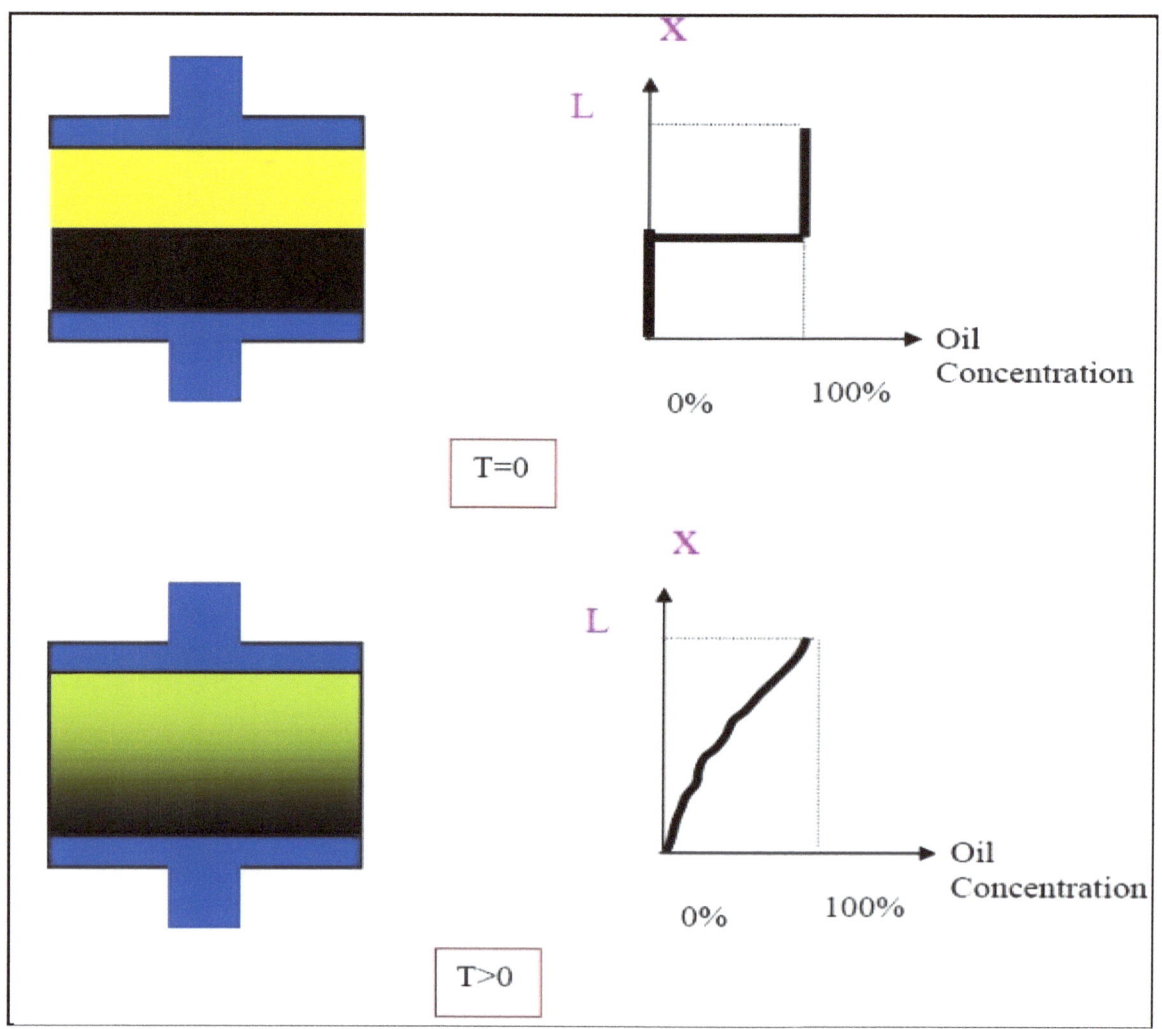

Figure 9: Diagram Illustrating the Diffusion Process within the DSR (Lesueur et al, 2005; and Walter et al, 2008)

Meanwhile, Karlsson and Isacsson (2003) opined that the major factors which facilitate an interaction between old and new binders in recycled mixtures are:

- mechanical mixing,
- compatibility (i.e. solubility parameter and molecular weight distributions) and
- diffusion

They further opined that the effectiveness of mechanical mixing is influenced by many factors i.e. temperature, binder viscosity, mixing time and type of mixture, and that compatibility between binders is a requirement for creating a homogeneous binder and is mainly dependent on the nature and distribution of intermolecular associations, which in turn determine the structural stability of binder. Also, they remarked that diffusion is a function of: temperature, intermolecular forces, structural rigidity of the diffusing molecules (restrictions on bending and twisting), microscopic structure of a relatively stationary phase if any, size and shape of the diffusing molecules or agglomerations as well as the viscosity of the medium in which the diffusion takes place. In conclusion, they submitted that the possibility of diffusion may not hold for low temperatures or severely aged binders and binders showing different behaviour under the influence of ageing.

In summary, it is obvious that more studies should be undertaken to ascertain the interaction between aged and virgin (rejuvenating) binders in cold recycled mixtures as information regarding this is still ambiguous. Though it has been established in the SCORE project that an interaction exists in such circumstances, the fact that the process involved is a time and temperature dependent physico-chemical one, and coupled with the fact that pavement engineers would normally desire that pavements start performing their roles efficiently shortly after being laid make it incumbent on researchers to put in place clear cut guidelines for the use of RAP in cold mixtures, which are based on a better understanding of the behaviour/interaction of the binders embedded in such cold recycled mixtures. For example, the degree of severity of ageing of binders in RAPs to be used in cold mixtures should be set in order to ultimately get good mixture performance almost immediately after construction.

12   Cold Mixes and Structural Design

Structural design is not completely new in cold mix design although there is no universally accepted method. This aspect of design is very important for determining (by optimisation) the layer thicknesses that should give the structural support required for the design life of a pavement. Thom (2009) suggested two general approaches for pavements incorporating cold mixture:

- treat it as hot-mix ,
- treat it as a very superior granular material.

Similarly, some methods which are basically empirical in approach have been advanced. These have been reported by Jitareekul (2009) as, the Catalogue Method, the CBR Design Method, the Asphalt Institute Method, AASHTO Method, and a UK method developed by Milton and Earland (1999).

In recent times also, the use of modelling tools involving multi layer linear elastic analysis which are commonly used for optimising layer thicknesses in pavements containing HMA have also been found useful for pavements which incorporate cold layers (Jitareekul, 2009; Ebels, 2008; and Twagira, 2010). These are implemented based on the results of performance tests conducted on mixtures. BISAR 3.0 and KENLAYER among other modelling tools have been chiefly used in this regard. Other methods which have been suggested are finite element modelling due to the complex nature of cold mixtures (Ebels, 2008; and Twagira, 2010), though this could be time consuming.

However, as much as such modelling tools are good for layer thickness optimisation, they must be executed with caution since cold mixtures have been found to be stress dependent i.e. non-linear in response (Ibrahim, 1998; Sunarjono, 2008; Ebels, 2008; and Twagira 2010) compared to HMA, whereas most such tools assume linear elastic behaviour.

13   Summary

The following are the conclusions from the review conducted in this chapter:

- Ageing of bitumen whether in the short or long term is inevitable in bituminous road pavements.
- Ageing in bituminous materials is evidenced by physical hardening which could either be advantageous or otherwise in a road pavement.
- Laboratory protocols (accelerated) for simulating ageing of bituminous road pavements in temperate climates have been developed. Such protocols are useful for predicting performance in service of bituminous road pavements.

- There is a need to develop similar laboratory protocols for road pavements in hot tropical climates.
- Cold mixes are cost effective, energy efficient and environmentally friendly, although inhibiting factors such as low early life stiffness and strength development, high air void contents and presence of moisture in the mix are still preventing such mixtures from being fully embraced in some developed countries.
- Most of the studies reported in literature on cold mixes have focused on temperate climates, though with good results in most cases.
- There is lack of evidence in literature of similar studies conducted in hot tropical climates and therefore there is a need to urgently look at this area.
- There is no universally accepted method for cold mix designs (job formula) although most of the methods which have been advanced in literature are based on the Hveem and Marshall methods of mix design.
- Adequate mixing and compaction alike are important for cold mixes.
- Curing is required for cold mixes for maturation and development of stiffness. The use of OPC in such mixes aids evaporation of moisture and achieves quicker stiffness and strength development.
- Stiffness, fatigue response, deformation properties, resilient modulus, and water susceptibility are good means for assessing the performance of cold mixes.
- Knowledge about the interaction between virgin binders and aged residual binders in cold recycled mixes is still ambiguous and needs further exploration. RAPs are still regarded as black rocks when they are used in cold recycling.
- The effect of preparation temperature (mixing and compacting) of cold mixes has not been fully accounted for in literature.
- Structural design of pavements involving cold mixtures should consider the non linear behaviour of such mixtures for good results.

References

Airey, G. D. (1997) Rheological *Characteristics of Polymer Modified and Aged Bitumens*. PhD Thesis, University of Nottingham.

Airey, G.D. (2003) State of the Art Report on Ageing Test Methods for Bituminous Pavement Materials. *The International Journal of Pavement Engineering*, Vol. 4 (3) September, pp165-176.

Airey, G. D., Choi, Y. K., Collop, A. C., Moore, A. J. V. and Elliot, R. C. (2005) Combined Laboratory Ageing/Moisture Sensitivity Assessment of High Modulus Base Asphalt

Mixtures. *Journal of the Association of Asphalt Paving Technologists*, Vol. 74, AAPT, White Bear Lake, pp. 307-341.

Artamendi, I. and Khalid, H. A. (2006) Diffusion Kinetics of Bitumen into Waste Tyre Rubber. *Journal of the Association of Asphalt Paving Technologists*, Vol. 75, AAPT, White Bear Lake, pp133-160.

Asphalt Institute and Asphalt Emulsion Manufacturers Association (1997) *A Basic Asphalt Emulsion Manual*. Manual Series No. 19, Third Edition, Asphalt Institute and Asphalt Emulsion Manufacturers Association, USA.

Bearsley, S., Forbes, A. and Haverkamp, R.G. (2004) Direct Observation of the Asphaltene Structure in Paving-grade Bitumen using Confocal Laser Scanning Microscopy. *Journal of Microscopy*, Vol. 182, Pt 1, April 1996, The Royal Microscopical Society.

Bell, C.A. (1989) *Summary Report on Aging of Asphalt-Aggregate Systems*. SHR-A/IR-89-004, Strategic Highway Report Research program, National Research Council.

Bell, C. A., Fellin, M. J. and Wieder, A. (1994) Field Validation of Laboratory Aging Procedures for Asphalt Aggregate Mixtures. *Journal of the Association of Asphalt Paving Technologists*, Vol. 63, AAPT, White Bear Lake, pp. 45-72.

Biczysko, S. J. (1996) Performance of Cold Recycled Bituminous Material. *Performance and Durability of Bituminous Materials*, (eds Cabrera, J. G. and Dixion, J. R.), E & FN Spon, London.

Bocci, M., Virgili, A. and Colgrande, S. (2002) A study of the Mechanical Characteristics of Cold Recycled Bituminous Concretes. *Proceeding of 4th European Symposium on Performance of Bituminous and Hydraulic Materials in Pavement, BITMAT 4*, University of Nottingham, UK, 11-12 April 2002, A.A. Balkema Publishers, Netherlands, pp 227-235.

Brennan, M.J., Gilbin, P.M., Kavanagh, A. and Sheahan, J. (2007) Laboratory Performance of an Emulsion-Bound Macadam Manufactured using different Processes. *Asphalt Professional* No. 29, November 2007, Institute of Asphalt Technology, 19-24.

Brown, S.F. and Needham, D. (2000) A Study of Cement Modified Bitumen Emulsion Mixtures. *Journal of the Association of Asphalt Paving Technologists*, Vol. 69, AAPT, White Bear Lake, pp 92-116.

Carswell, J. (2004) *SCI: The Road from Vienna: Eurobitumen/EAPA Conference Themes*. (Summary of Moderator Reports), www.soci.org/SCI/groups/cmt/2004/reports/pdf/gs-3072.pdf (Accessed: 10/05/2008).

Carswell, J., Ellis, S. J. and Hewitt, A. (2008) Design and Specification for Sustainable Maintenance of Roads using Cold Recycling Techniques. *Review of the Growth and Development of Recycling in Pavement Construction*, World Road Association (PIARC), Cedex, pp 169-181.

Chen, J.S., Huang, C.C. and Lin, K.Y. (2007) Engineering Characterization of Recycled Asphalt Concrete and aged Bitumen Mixed Recycling Agent. *Journal of Material Science*, Vol. 42, No. 23, December, 2007, Springer Netherlands, http://www.springerlink.com/content/r443650h6l473716/ (Accessed: 10/05/2010).

Ebels, L-J. (2008) *Characterisation of Material Properties and Behaviour of Cold Bituminous Mixtures for Road Pavements*. PhD Dissertation, Stellenbosch University.

Epps, J. A., Little, D. N., Holmgreen, R. J. and Terrel, R. L. (1980) *Guidelines for Recycling Pavement Materials*. NCHRP 224, National Research Council, Washington, DC.

FHWA (1997) *Pavement Recycling Guidelines for State and Local Governments*. Participant's Reference Book, Federal Highway Administration, USA, http://www.fhwa.dot.gov/pavement/recycling/98042/ (Accessed: 20/11/2007).

GEOPAVE (1993) *Foam Bitumen Stabilised Pavements*. GEOPAVE Materials Technology Technical Note 8, http://webapps.vicroads.vic.gov.au/VRNE/vrbscat.nsf/e5ff054ca38faf2b052568550077d3 e7/9c0d4ec7447adf53ca256c6f0019c484/$FILE/tn008.pdf (Accessed: 20/12/2007).

Hachiya, Y., Nomura, K. and Shen, J. (2003) Accelerated Aging for Asphalt Concretes. *Proceedings of the 6th International RILEM Symposium on Performance testing and Evaluation of Bituminous Materials*, Zurich, (ed. Partl, M.N.), RILEM Publications, Bagneux, pp 133-140.

Harun, M.H. and Morosiuk, G (1995) *A Study of the Performance of Various Bituminous Surfacings for Use on Climbing Lanes*. Overseas Centre, TRL, Crowthorne.

Hinds, S. (2007) *Re-recycling Asphalt*. M.Eng. Report, University of Nottingham.

Ibrahim, H. E. and Thom, N. H. (1997) The Effect of Emulsion-Aggregate Mixture Stiffness on both Mixture and Pavement Design. *Proceedings of the Second European Symposium on Performance and Durability of Bituminous Materials*, Leeds, pp351-367.

Ibrahim, H.E.M. (1998) *Assessment and design of Emulsion Aggregate Mixtures for Use in Pavements*. Ph.D. Thesis, University of Nottingham.

Jenkins, K. J. (2000) *Mix Design Considerations for Cold and Half-Warm Bituminous Mixes with Emphasis on Foamed Bitumen*. PhD Thesis, University of Stellenbosch.

Jitareekul, P. (2009) *An Investigation into Cold In-Place Recycling of Asphalt Pavements*. PhD Thesis, University of Nottingham.

Jitareekul, P., Sunarjono, S., Zoorob, S.E. and Thom, N.H. (2007) Early Life Performance of Cement and Foamed Bitumen Stabilized Reclaimed Asphalt Pavement under Simulated Trafficking. *Proceedings of Special Sessions, Sustainable Construction Materials and Technologies*, Coventry, pp. 308-321.

Jove, M.B. and Bock, L.D. (2003) *The PARAMIX Project: Enhanced Recycling Techniques for Asphalt Pavements*. www.cimne.com/paramix/ppt/263_Wascon.doc (Accessed: 20/11/2007).

Judycki, J. and Jaskula, P. (2003) Testing of Performance Properties of Asphalt Mixes for Thin Wearing Courses. *Proceedings of the 6th International RILEM Symposium on Performance testing and Evaluation of Bituminous Materials*, Zurich, (ed. Partl, M.N.), RILEM Publications, Bagneux, pp 141-147.

Karlsson, R. (2002) *Investigations of Binder Rejuvenation Related to Asphalt Recycling*. PhD Thesis, Division of Highway Engineering, Royal Institute of Technology, Stockholm.

Karlsson, R. and Isacsson, U. (2003) Application of FTIR-ATR to Characterization of Bitumen Rejuvenator Diffusion. *Journal of Materials in Civil Engineering*, Vol. 15, No. 2, ASCE, pp. 157-165.

Khalid, H. A. (2003) Assessing the Potential in Fatigue of a Dense Wearing Course Emulsified Bitumen Macadam. *Proceedings of the 6th International RILEM Symposium on Performance testing and Evaluation of Bituminous Materials*, Zurich, (ed. by Partl, M.N.), RILEM Publications, Bagneux, pp 349-356.

Khalid, H. and Eta, K. E. (1997) Structural Support Values for Emulsified Bitumen Macadams in Highway Reinstatement. *Proceedings of the Second European Symposium on Performance and Durability of Bituminous Materials*, Leeds, pp 307-336.

Kim, K. W. and Burati, J. L. (1993) Use of GPC chromatograms to characterize aged asphalt cements. *Journal of Materials Engineering in Civil Engineering*, Vol. 5, No. 1, ASCE, Reston, pp 41-52.

Kim, Y. and Lee, H.D. (2006) Development of Mix Design Procedure for Cold In-Place Recycling with Foamed Asphalt. *Journal of Materials Engineering in Civil Engineering*, Vol. 17, No. 5, January/February, 2006, ASCE.

Kim, Y., Lee, H.D., and Heitzman, M. (2007) Validation of New Mix Design Procedure for Cold In-Place Recycling with Foamed Bitumen. *Journal of Materials Engineering in Civil Engineering*, Vol. 17, No. 5, November 1, ASCE.

Kulash, D.J. (1994) Toward Performance-Based Specifications for Bitumen and Asphalt Mixtures. Paper 10190, *Proc. Instn. Civ. Engrs Transp.*, 1994, 105, Aug, 187-194.

Lee, H.D. (2007) *Development of Mix Design process for Cold In-Place Recycling Using Emulsion*. A Proposal Presented to Iowa Highway Research Board, Phase III.

Lee, T-C, Terrel, R. and Mahoney, J. (1983) *Test for Efficiency of Mixing of Recycled Asphalt Paving Mixtures*. Transportation Research Record 911, Transportation Research Board, National Research Council, Washington, DC, pp 51-66.

Leech, D. (1994) *Cold Bituminous Materials for use in the structural layers of roads*. Transport Research Laboratory Project Report 75, Overseas Centre, TRL, Crowthorne.

Lesueur, D., Potti, J. J., Southwell, C., Walter, J., Lancaster, I., Cruz, M., Delfoss, F., Eckmann, B., Maze, M., Fiedler, J., Racek, I., Simonsson, B., Brosseaud, Y., Serrano, J., Ruiz, A., Kalaaji, A. and Attane, P. (2005). *The SCORE Project: Superior Cold Recycling*.

Loizos, A. and Papavasiliou, V. (2006) Evaluation of Foamed Asphalt Cold in-Place Pavement Recycling Using Non-destructive Techniques. *Journal of Transportation Engineering*, vol. 132, No. 12, December 1, 2006, ASCE.

Long, F. and Theyse, H. (2004) Mechanistic-Empirical Structural Design Models for Foamed and Emulsified Bitumen Treated Materials. *Proceedings of the 8th Conference on Asphalt Pavements for Southern Africa (CAPSA'04)*, Sun City.

Lu, X. and Isacsson, U. (2002) Effect of Ageing on Bitumen Chemistry and Rheology. *Construction and Building Materials*, 16 (2002), Elsevier Science Ltd, pp 15-22.

Mastrofini, D. and Scarsella, M. (1999) The Application of Rheology to the Evaluation of Bitumen Ageing. *Fuel* 79, Elsevier Science Ltd, pp1005-1015.

McDaniel, R. S., Soleymani, H., Anderson, R. M., Turner, P. and Peterson, R. (2000) *Recommended use of Reclaimed Asphalt Pavement in the Superpave Mix Design Method.* Web Document No. 30 (Project No. D9-12): Contractor's Final Rep., NCHRP.

Milton, L. J. and Earland, M. G. (1999) *Design Guide and Specification for Structural Maintenance of Highway Pavements by Cold In-Situ Recycling.* TRL Report TRL386, Transportation Research Laboratory, Crowthorne.

Montepara, A. and Giuliani, F. (2002) A Study on Design and performance of Recycled Pavement Cold Stabilized with Cement and Bituminous Emulsion. *Proceeding of 4th European Symposium on Performance of Bituminous and Hydraulic Materials in Pavement,* BITMAT 4, University of Nottingham, UK, 11-12 April 2002, A.A. Balkema Publishers, Netherlands, pp 213-217.

Mouillet, V., Lamontagne, J., Kister, J. and Martin, D. (2003) Development of a New Methodology for Characterization of Polymer Modified Bitumens Ageing by Infrared Microspectrometry Imaging. *Proceedings of the 6th International RILEM Symposium on Performance testing and Evaluation of Bituminous Materials,* Zurich, (ed. Partl, M.N.), RILEM Publications, Bagneux, pp 153-159.

Needham, D. (1996) *Developments in bitumen emulsion mixtures for roads.* PhD Thesis, University of Nottingham.

Nguyen, V. H. (2009) *Effects of Laboratory Mixing Methods and RAP Materials on Performance of Hot Recycled Asphalt Mixtures.* PhD Thesis, University of Nottingham.

Noureldin, A. S. and Wood, L. E. (1987) *Rejuvenator Diffusion in Binder Film for Hot-Mix Recycled Asphalt Pavement.* Transportation Research Record 1115, Transportation Research Board, National Research Council, Washington, DC, pp 51.

Oke, O. L., Parry, T, Thom, N and Airey, G. A. (2010), *Laboratory Ageing Protocols for Asphalt Recycling in Hot Climates,* Proceedings of Special Technical Sessions, Second International Conference on Sustainable Construction Materials and Technologies June 28 - June 30, 2010, Università Politecnica delle Marche, Ancona, Italy. Pg 291-302.

Petersen, J. C. (1984) *Chemical Composition of Asphalt as Related to Asphalt Durability: State of the Art.* TRR 999, Transportation Research Board, Washington, DC, pp 13-30.

Petersen, J. C., Barbour, F. A. and Dorrence, S. M. (1974) Catalysis of Asphalt Oxidation by Mineral aggregate Surface and Asphalt Components. *Proc. Association of Asphalt Paving Technologists,* Vol. 43, pp 162-177.

PIARC (2003) *Pavement Recycling Guidelines.* World Road Association (PIARC), Cedex.

Roberts, F. L., Engelbrecht, J. C. and Kennedy, T. W. (1984) *Evaluation of Recycled Mixtures Using Foamed Asphalt.* Transportation Research Record No. 968.

Roberts, F. L., Kandhal, P. S., Brown, E. R., Lee, D. Y. and Kennedy, T. W. (1991) *Hot Mix Asphalt Materials, Mixture Design, and Construction.* NAPA Education Foundation, Maryland.

Romanoschi, S.A., Heitzman, M. and Gisi, A.J. (2003) Foamed Asphalt Stabilized Reclaimed Asphalt: A Promising Technology for Mid-Western Roads. *Proceedings of the August 2003 Mid-Continent Transportation Research Symposium,* Ames, Iowa.

Romera, R., Santamaria, A., Pena, J.J., Munoz, M.E., Barral, M. Garcia, E. and Janez, V. (2006) *Rheological Aspects of the Rejuvenation of Aged Bitumen*. Rheol Acta, Springer-Verlag, pp. 474-478.

Santucci, L. E. (1977) Thickness Design Procedure for Asphalt and Emulsified Asphalt Mixes. *4th Int. Conf. - Ann Arbor*, pp 424-456.

Schiavi, I., Nunn, M., Nicholls, C. and Chambers, P. (2003) Effectiveness and durability of Rejuvenating Agents. *Proceedings of the 6th International RILEM Symposium on Performance testing and Evaluation of Bituminous Materials*, Zurich, (ed. Partl, M.N.), RILEM Publications, Bagneux, pp148-152.

Scholz, T.V. (1995) *Durability of Bituminous Paving Mixtures*. PhD Thesis, School of Civil Engineering, University of Nottingham.

Serfass, J-P., Poirier, J-E., Henrat, J-P. and Carbonneau, X. (2003) Influence of Curing on Cold Mix Mechanical Performance. *Proceedings of the 6th International RILEM Symposium on Performance testing and Evaluation of Bituminous Materials*, Zurich, (ed. Partl, M.N.), RILEM Publications, Bagneux, pp 81-87.

Shen, J., Amirkhanian, S. N. and Lee, S-J. (2007) HP-GPC Characterization of Rejuvenated Aged CRM Binders. *Journal of Materials in Civil Engineering*, Vol. 19, No. 6, June 1, 2007, ASCE, Reston.

Smith, H.R. and Jones, C.R. (1998) *Bituminous Surfacings for Heavily Trafficked Roads in Tropical Climates*. Paper 11513, *Proc. Instn. Civ. Engrs Transp.*, 1998, 129, Feb, ICE, 28-33.

Soenen, H., Sandman, B. and Nilsson, A. (2000) Rheological and Chemical Evaluation of the Ageing of SBS Modified Bitumen as Used in Roofing. *11th International Congress Proceedings*, International Waterproofing Association, Nottingham.

Soleymani, H.R., Bahia, H.U. and Bergan, A.T. (1999) Time-Temperature of Blended Rejuvenated Asphalt Binders. *Journal of the Association of Asphalt Paving Technologists*, Vol. 68, AAPT, White Bear Lake, pp. 129.

Srivastava, A. and Rooijen, R.V. (2000) Bitumen Performance and Arid Climates. *Pavement Seminar for the Middle East and North Africa Region: Innovative Road Rehabilitation and Recycling Technologies, New Polices and Practices in Pavement Design and Execution*, Amman.

Staple, P.R. (1997) Cold Emulsion Macadam Performance Trials for Footway Surfacing in Leicestershire. Paper 11298, *Proc. Instn. Civ. Engrs Transp.*, 1997, 123, Aug, 174-177.

Sunarjono, M. T. (2008) *The Influence of Foamed Bitumen Characteristics on Cold-Mix Asphalt Properties*. PhD Thesis, University of Nottingham.

Thanaya, I. N. A. (2003) *Improving The Performance of Cold Bituminous Emulsion Mixtures (CBEMs) Incorporating Waste Materials*. PhD Thesis, School of Civil Engineering, the University of Leeds, Leeds.

Thanaya, I. N. A. (2007a) Evaluating and Improving the Performance of Cold Asphalt Emulsions Mixes, *Civil Engineering Dimension*, September 2007, Vol. 9, No. 2, 64-69, Indonesia.

Thanaya, N.A. (2007) Review and Recommendation of Cold Asphalt Emulsion Mixtures Design. *Civil Engineering Dimension*, March 2007, Vol. 9, No. 1, Indonesia.

Thom, N. H. (2008) *Principles of Pavement Engineering.* Thomas Telford Publishing Limited, London.

Thom, N. H. (2009) *Cold-Mix Asphalt.* Lecture Notes, University of Nottingham.

Traxler, R. N. (1963) *Durability of Asphalt Cements. Association of Asphalt Paving Technologists*, Vol. 32, pp 44-63.

Twagira, E. M. (2010) *Influence of Durability Properties on Performance of Bitumen Stabilized Materials.* PhD Dissertation, Stellenbosch University.

Ulmgren, N. (2003) Gyratory Compaction- Influence of Compaction angle on Stability and Stiffness Characteristics. *Proceedings of the 6th International RILEM Symposium on Performance testing and Evaluation of Bituminous Materials*, Zurich, (ed. Partl, M.N.), RILEM Publications, Bagneux, pp 244-249.

Walter, J., Attané, P., Kalaaji, A. and Lancaster, I. (2008) Regeneración del betún en el proceso de reciclado en frío (Bitumen regeneration in cold recycling processes). *Carreteras*, número 158/Mar-Abr 08, pp 27-35.

Walubita, L.F., Martin, A.E., Glover, C., Jung, S.H., Cleveland, G. and Lytton, R.L. (2005) *Fatigue Characterization of HMAC Mixtures Using Mechanistic Empirical and Calibrated Mechanistic Approaches including the Effects of Aging, Asphalt Concrete: simulation, Modeling, and Experimental Characterization.* Proceedings of the R. Lytton Symposium on Mechanics of Flexible Pavements, Geotechnical Special Publication No. 146, ASCE, pp. 103-114.

Welborn, J. Y. (1979) *Relationship of Asphalt cement properties to pavement durability.* NCHRP Synthesis 59, Washington, DC.

Widyatmoko, I., Elliot, R.C., Heslop, M.W. and Williams, J.T. (2002) Ageing Characteristics of Some Low Penetration Grade Binders. *4th European Symposium on the Performance of Bituminous and Hydraulic Materials in Pavements*, University of Nottingham.

Zoorob, S. E. and Thanaya, I.N.A. (2002) Improving the Performance of Cold Bituminous Emulsion Mixtures (CBEMs) Incorporating Waste Materials. *Proceeding of 4th European Symposium on Performance of Bituminous and Hydraulic Materials in Pavement, BITMAT 4, University of Nottingham, UK, 11-12 April 2002*, A.A. Balkema Publishers, Netherlands, pp 237-249.

Zube, E. and Skog, J. (1969) *Final Report on the Zaca-Wigmore Asphalt Road Test.* AAPT, Vol. 38, pp 1-38.

www.ingramcontent.com/pod-product-compliance
Lightning Source LLC
Chambersburg PA
CBHW040746200526
45159CB00023B/1748